A M
OF ITS
OWN?

ChatGPT and the Surprising World of the New AI

Terry M. Savage

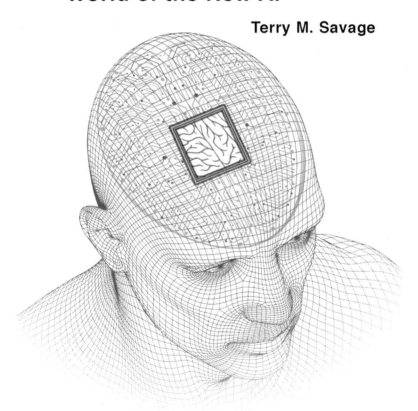

2025

Brentwood, New Hampshire

Savage, Terry M.

A Mind of Its Own?
ChatGPT and the Surprising World of the New AI

TerryMSavage@gmail.com

ISBN: 979-8-218-57942-5
Library of Congress Control Number: 2025900029

Book and cover design by Robin Wrighton

For Jane,

a thoughtful partner in this project,

as in so much else.

Contents

Introduction

The release of ChatGPT in the fall of 2022 created a public sensation.

Sensationalism was nothing new to *Artificial Intelligence* or AI, the field of computer science devoted to creating, in one form or another, a "thinking machine." Promises of the imminent arrival of devices that would duplicate human intelligence were quite literally as old as the 70-year career of AI itself. Somehow, the field always managed to disappoint: the latest "breakthrough" never translated to a more general, and convincing, machine intelligence.

AI had its share of successes, of course, and many of these had found their way into daily life. We have long grown accustomed to machines that could read our handwriting, or generate speech from the text we typed, or type text from the speech we spoke.

We can also speak commands to set our cooking timers and play our favorite songs. We can talk to machines and machines will talk back to us. This is helpful, and it also marks a significant accomplishment in "natural language processing" (NLP), the area of AI focused on enabling computers to converse with users in English and other natural languages.

ChatGPT had emerged from advanced research in NLP and it was more or less to be expected that it would improve the fluency of "talking" machines. In fact, it spoke quite well; but after years of conversing with Siri, Alexa, and other digital assistants another talking machine was hardly surprising let alone sensational.

The surprise, and the sensationalism, had two sources. The first was that ChatGPT didn't simply talk; it also seemed to *know what it was talking about.* Unlike a Siri or an Alexa, ChatGPT could enter into a meaningful conversation on virtually any topic. It had apparently discovered not only the rules of effective language use, but also the elements of knowledge expressed in that language. This was impressive and quite noteworthy.

The second surprise, however, was more than impressive; it was simply shocking. The shock lay in what ChatGPT could *do* with the knowledge that it had discovered. ChatGPT didn't just speak English and know a lot about what humans had written; it could summarize articles and books, it could translate languages, do math, write computer code, diagnose illnesses, prepare legal briefs, and pass any number of exams devised to test the knowledge and skills of humans. It was, in fact, difficult to find an intellectual task that it could not perform, at least at some level.

As a "Large Language Model" (LLM), ChatGPT had been built to perform a rather basic task: predict the next likely word in a given sequence of words. Given, for instance, "The dog heard the noise, so it _____," the model might learn to generate "barked" as its prediction. The model was, to be sure, complex, sophisticated, and powerful. But it had only been told to predict next words; no one had told it how to build on its predictions to develop its impressive range of abilities.

The abilities had simply *emerged* from the model itself, once it had reached a certain size. AI, this time, had not over-promised and under-delivered. It had done quite the opposite, surprising even the scientists and engineers who, quite unintentionally, had under-promised and over-delivered. The machine surprised nearly everyone by developing something much like knowledge, and even thought, *all on its own.*

Had GPT found, by itself, a path to intelligence? Were we now in the presence of a new powerful device with, in some sense, a "mind" of its own?

Not everyone thinks so. Some have dismissed ChatGPT, and other LLMs, as "glorified auto-completion" programs. Other critics have described them as giant guessing machines that produce their output, not "knowingly," but solely through statistical probabilities.

To the challenges of the critics who question the "intelligence" of LLMs, we can add the more worldly doubts and cautions of some economists and corporate executives. The splash made by ChatGPT and other LLMs was followed by a virtual stampede of investment in the technology. NVIDIA, the leading provider of the computer chips essential to the development of language models, reached a market value of approximately three *trillion* dollars by the fall of 2024. This was larger, as one source puts it, than the entire economies of some nations.

But would the enthusiasm hold? Would the new technology so transform the world that these huge investments would produce an actual return on investment? Is the New AI a genuinely "intelligent" technology likely to insert itself into every corner of our lives, transforming our economies and even our broader social relationships?

Today, we are all, AI scientists and engineers included, spectators in an unfolding drama. With a mixture of awe and trepidation, we are, perhaps, watching the arrival of a new intelligence among us.

Or, are we watching the greatest of all the cycles of AI hype? As one distinguished scientist has argued (see Part Four), language models speak so well that they fool us into believing that some form of thought must lie behind their speech. Humans have often been quick to attribute their own characteristics to non-humans. Perhaps the so-called intelligence of an LLM is just a projection by humans of their own intelligence, a naïve assumption that the thought that accompanies our own speech must also be present in the machine.

There is no shortage of enthusiasm, doubt, or debate in the various reactions to the arrival of ChatGPT. How should we judge the arguments? How do we make sense of this variant of "machine intelligence?"

The five parts of the study that follows are intended as an orientation for those who would observe, perhaps a bit more closely, the drama unfolding before us. In the first, we explore the reasons for the current "fuss over AI." In the second, we peer into the "black box" of ChatGPT and language models in general to determine how they work. The third explores the capabilities and limitations of LLMs and the fourth seeks to determine the nature of LLM "intelligence." The fifth, finally, offers some thoughts on the likely future of "living with the new AI."

The arrival of a new technology is inevitably disruptive. This one has already generated more than a little worry: Will it be used by bad actors for social and political manipulation, for fraud and other crimes? Can it be used to produce new chemical or biological weapons? Will it take the jobs of millions of "knowledge workers?" Has it already irreversibly undermined education by offering such a tempting substitute for the hard work of learning and writing?

Full disclosure: As the last worry may suggest, I come, in part, from an academic background. For many years, I taught a college course focused on the technology and the social impact of artificial intelligence. As a result, I had a passing familiarity with the history and accomplishments of the field.

My reaction to ChatGPT was immediate: "No, it can't! It should not be able to do this!" I knew, on a global level at least, the different strategies that had been used to embody "intelligence" in a computing machine. The problems that had, apparently, been solved by ChatGPT were very challenging and attempts to master them had often ended in frustration and failure.

How, on earth, had a "next word predictor" solved them? What other problems might it be able to solve?

Over the course of a year or so, I have tried to understand just what this new technology actually does and what it may mean for all of us who are going to be living with it. I was motivated by surprise and puzzlement. What follows are the answers, many provisional, and the still unresolved questions, sparked by the arrival of ChatGPT.

A final disclosure, occasioned by the technology itself: No, ChatGPT has not written a single sentence of this study. Writing, especially for the puzzled and curious, is not about recording what you know; it's about learning what you do not know. No one should *delegate* that – not to another person, and certainly not to a machine.

But did ChatGPT contribute to this little book? Most definitely, and in a manner that, for me at least, was a bit transformative. ChatGPT (in the form of GPT-4o) has been at my shoulder, a *semi-*trusted research assistant always readily available to answer a question or review my writing for accuracy. Yes, semi-trusted; no one should uncritically accept the output of an LLM. That said, I would not now want to be without the assistance a robust LLM has to offer.

My time with ChatGPT, Claude, and other LLMs has left me quite aware of their limitations but continually surprised by their capabilities. Others have also been surprised.

Perhaps we should try to understand just what they are and what they might become.

Part One

The Fuss Over AI

All of a sudden, many people are very worried about the development of artificial intelligence. In Europe, legislators have enacted guidelines and restrictions. In the U.S., leaders within the AI research community called for a moratorium on the development of their own technology. The U.S. Congress has held hearings. The U.S. president has issued a sweeping Executive Order in an attempt to guide and regulate the advance of AI. The promise and threat of a "revolutionary" technology are headline news.

Why all the fuss?

There is a short answer. On November 30, 2022, OpenAI, an artificial intelligence research company, launched ChatGPT. ChatGPT was based on the Large Language Model (LLM) GPT-3 and was specially configured to respond to user questions and requests expressed in ordinary English. OpenAI made access to ChatGPT free and readily available.

Within two months of its launch, ChatGPT had set a new record in attracting 100 million users per month. Tik Tok had taken nine months to reach 100 million users; Instagram took 2.5 years. As of November, 2023, ChatGPT had over 100 million active users *each week*.[1]

ChatGPT is not the only successful LLM; but it is arguably the best known and the most widely discussed. This "language model" surprised, and even shocked, many of its users. It had been developed by OpenAI to facilitate communication between humans and machines in natural language (English, French, Mandarin, etc). We can write, or speak, a request in English, or another language, and ChatGPT will answer very much as another human would. It is also, in fact, quite "natural" in its use of language. The shock is not that it speaks so well. The shock is in what it has to say and what it can do.

ChatGPT seems to have mastered much more than the ability to converse in ordinary language (so-called NLP, Natural Language Processing). Its hundreds of millions of users have discovered that it can summarize books and articles, write poetry and short stories, plan travel itineraries, perform mathematical calculations, write computer code, translate languages, diagnose illnesses, pass a wide range of competency tasks in areas such as law and college admissions, and much more. A general public accustomed to splash AI triumphs in limited, specialized areas (Chess, Go) and only marginally aware of behind-the-scenes implementations in areas such as insurance, banking, or social media was suddenly presented with a seemingly multi-purpose intelligence.

ChatGPT, as a sort of "point of the spear," made the world aware of a demonstrably powerful technology with the potential to affect virtually everyone in one way or another. Users were repeatedly surprised by its apparent mastery of such a wide range of tasks.

The surprise, moreover, was not confined to the general public. Developers programmed ChatGPT to have conversations by predicting the next word in a sequence of words. They did not program it to perform any of the wide range of additional tasks of which it is capable. Instead, these are "emergent behaviors" learned by the models themselves; and their appearance surprised even their creators. There was something quite unusual about a program developed to speak in English that could also instantly respond in French or write a computer program to perform sales tracking in C (or virtually any other programming language).

To understand just why ChatGPT's capabilities have produced such a fuss, a bit of historical context is useful.

AI and Two Paths to Machine Intelligence

The beginnings of artificial intelligence are generally traced to a conference held on the campus of Dartmouth College in the summer of 1956. It was there that the term was first introduced; and it was there that some of the first seeds of the practical work of building intelligent machines were planted.

In an important sense, however, the modern dream of a thinking machine begins twenty years earlier. In 1936, Alan Turing published a paper that introduced the world to a creation that would come to be known as a Universal Turing Machine.[2]

A Universal Machine

Turing had not set out to devise a new kind of machine. He was a mathematician and he created his machine as a way of giving the abstract notion of an "effective procedure" a clear and concise definition. In this, he was successful. He first described an abstract machine to process information and especially mathematical information. He then showed how to create a machine that would be able to duplicate the functions of that machine and, critically, *any other* conceivable information-processing machine. The latter was a "Universal Turing Machine" (UTM). Today, we call such a machine a *computer.* Turing had laid the foundations for the computer revolution.

Turing's machine was an abstraction, a sort of thought experiment, offered as a way to answer a certain mathematical question.[3] It demonstrated the theoretical possibility of computing machinery without itself being a physical machine. Early computers, the physical machines that embodied the underlying principles of UTMs, were also focused on mathematics. Indeed to most early users, computers were for "computing" in the narrow sense of performing mathematical computations. They offered obvious advantages in a wide range of applications in which numbers and their manipulation were central.[4] Early computers were easily embraced as an extension of the mechanical and electromechanical calculators already in widespread use. The computer, from this perspective, was just another clever, special-purpose machine.

Turing, however, seems to have almost immediately recognized the implications of what was a fundamentally different sort of machine. And we can readily forgive those early adopters for their limited vision: *all* previous machines were, in fact, special-purpose devices. They were built for a certain range of tasks. Outside that range, one needed to build a different machine. Turing, on the contrary, had shown that his machine could perform *any* information-processing procedure and thus control any other kind of machine. This made the UTM, theoretically at least, capable of doing virtually anything that can be done by any other device.[5]

And, to Turing, this meant that his machine should be able to do what a brain can do. It should be capable of thought itself.

In 1936, Turing had shown the world how to create a universal machine. In 1950, he provided a test to determine whether or not his machine, the computer, could actually *think*. In "Computing Machinery and Intelligence," he presented the test as the "imitation game."[6]

In a contemporary formulation of the "Turing Test," one person, as judge, communicates with two other participants via a text terminal. Thus, the judge cannot see or hear the interlocutors. One interlocutor is a person and the other is a computer. The judge's task is to determine which is the computer and which is the person by asking questions of each. The task of both interlocutors is to convince the judge that they are the person. If the judge cannot tell the difference, the computer wins; and if the computer wins we must concede that it is capable of thinking. Such a computer, Turing insisted, would be an *intelligent* machine.

Turing's test has been widely debated. It has been dismissed as inadequate by a number of commentators.[7] Some, for instance, have conceded that computers might effectively *mimic* the human use of language; but, they insist, machines cannot have the consciousness and awareness of mental states that underlies a human's *understanding* of language.

Many others consider a computer's ability to pass the test to be, at the very least, a significant milestone in the attempt to engineer a general-purpose, human-like intelligence (so-called AGI, Artificial General Intelligence). The test remains relevant today: some believe that LLMs have already passed it.

In the same article, Turing briefly considered the processes through which a computer might be endowed with human-like thinking. He noted that a part of the process might include giving the computer rules for playing games such as chess. He also suggested that certain rules of logical inference could be directly programmed thus providing the machine with a sort of core reasoning capability. In short, Turing recognized that there are aspects of intelligence that we can analyze and understand as sequences of steps that can then be represented in computer programs.

Symbolic AI

Many AI researchers have followed this general approach, which is often called "symbolic AI."[8] In symbolic AI, researchers duplicate an aspect of intelligence by creating a program or programs (a sequence of *symbols*) that capture their understanding of its essential properties. Thus, for instance, as Turing predicted, some early AI researchers focused on game playing. Many games eventually yielded to lines of code created by programmers as expressions of their understanding of the game's rules and strategy.

Early efforts also tackled general reasoning capabilities. In 1956, for instance, Allen Newell, Herbert Simon, and Cliff Shaw developed "Logic Theorist," a program that could prove mathematical theorems.[9]

Symbolic AI remains an important aspect of AI today; and it produced a number of noteworthy successes, such as the oft-cited victory of Deep Blue over the reigning chess champion, Gary Kasparov, in 1997. "Expert Systems" of the 1980s, which sought to capture, preserve, and make available the expertise of humans in a range of areas were also generally based on symbolic techniques.

In one of its most ambitious, and controversial, applications, symbolic AI was also applied to the problem of giving a computer "common sense." CYC, a programming project with the goal of directly encoding all human "common sense" knowledge began in 1984 and continues to this day.[10]

The successes of symbolic AI were a confirmation that there are certain aspects of intelligent operations that are achievable as a set of step-by-step instructions encoded in a computer program. We can, for instance, effectively encode many mathematical and logical operations. In such cases, researchers are, in effect, encoding their own knowledge, their own understanding.[11]

Turing's proposals, however, were not limited to an encoding of what programmers themselves know. He believed that computers could also be designed to *learn by themselves.* He envisioned a computer with a sort of "child brain" that rendered it capable of adding to its store of knowledge by experience of the world via a set of sensors. The key to the learning process was to be a system of rewards and punishments, reinforcing correct understandings and discouraging mistakes. These ideas, as he himself insisted, were merely sketches, suggestions of paths forward. The suggestion, however, relates very directly to a second major approach to AI research that many see as powerfully vindicated today.

Neural Networks and Connectionism

This second major AI strategy aimed to create a machine that could learn on its own. Research would focus on the problem of endowing computers with learning mechanisms. The basic mechanism they adopted was modeled, if only roughly, on the human brain.

As early as the 1940s researchers (notably Warren McCulloch and Walter Pitts) recognized that the fundamental element of a human brain, the neuron, could be partially represented on a Turing machine. The "perceptron" was a simulated artificial neuron. Perceptrons can be connected to one another to form an *artificial neural network.*

In the simplest version of a neural network input neurons are directly connected to output neurons. The input, for example, might be the image of a square, a circle, or some other shape. The task of the network might be simply to determine if the image it has been shown is, or is not, a circle. The network could have many input neurons, each of which presents, as its input, a specific part of the image. These "image parts" could be the individual "dots" or *pixels* of a digitized image represented by numerical values. These values (e.g., 0 for white and 1 for black in a black and white image) are given to the output neuron together with adjustable *weight* values that indicate the strength of the connection between the input and output neurons. The output neuron receives the values of input neurons and their weights, makes certain calculations, and gives its answer: either that the image is a circle or that it is not.

The answer, almost certainly, is wrong. What the network has completed is just the first of very many training sessions. Developers have randomly set the initial values of the network's weights. They do not know what those weights should be. The network, itself, will determine the values of the weights.

The training, in this case, uses images that are labeled. In effect, the network can be shown the label and learn that it is wrong. It can then recalculate the values of weights to reduce its error. It repeats the process many times for a series of images, some of which are circles while others are not. Repeated corrections gradually result in the network being able to correctly identify images as circles. The values that it has determined for the weights of connections between inputs and its output are, in effect, its record of the image patterns that define circles.

Such a machine is not *told how* to identify circles. It discovers how to do this by itself; and when it has determined the weight values that produce correct answers it has a form of "knowledge." The weights are a reflection of the patterns it has analyzed. It now "knows" the image patterns that define circles.

This approach to AI, based on neural networks, is known as *connectionism,* a name that reflects the role of neuron connections in machine learning. Connectionist research received early government support in the U.S. This allowed Frank Rosenblatt to develop his prototype "Mark 1 Perceptron," which was publicly demonstrated in 1958. Rosenblatt and the connectionists that followed him believed that the key to fully developed AI lay in teaching machines to learn on their own. A sort of competition soon developed between the two major schools of AI with both funding and scholarly reputations at stake. Symbolic AI doubted the capacity of neural networks to produce logical or mathematical inferences; connectionists believed that hand coding would fail to capture many essentials of human perception, limiting the reach of any intelligent machine.

Each approach gradually accumulated successes in relatively narrow areas; but each also failed to develop machines that could produce intelligent actions beyond the specific tasks for which they were designed. A key challenge was "common sense" reasoning. Computers, as it turned out, could be programmed to play expert chess, or prove mathematical theorems. They had much more trouble recognizing that cats don't play violins or that a person who is "over the moon" probably hasn't taken flight.

Turing had predicted in 1950 that, in fifty years time, a computer would perform well enough on his test to be judged intelligent. Interestingly, he thought that only a small increase in computing power would be required. The year 2000 came and passed with no indication that a computer could yet win the "Imitation Game."

And, then, things began to change in ways that vindicated the connectionist agenda. These changes in neural network computing will be explored in more detail in Part Two as we seek to understand how ChatGPT and other LLMs are able to do what they do. In broad terms, however, the past fifteen years have seen major developments in at least three key areas: learning model design, computing power, and data acquisition.

Early neural networks, such as Rosenblatt's Mark I, processed data through two layers: an *input layer* and an *output layer*. Later, in the 1980s, researchers added multiple *hidden layers* that carried out additional processing between inputs and outputs. By the 2010s advancements in learning model design had led to the possibility of employing tens of layers and the term "deep learning network" became commonplace. The "deepest" of today's learning models can have hundreds of layers.

Improvements in learning models also include a number of techniques for more efficient data entry and more efficient processing of that data. A number of these, including a particularly important research paper published in 2017, will be considered in Part Two.

There are also many factors involved in the increased computing power that propelled the success of deep learning networks. One of these was improved parallel processing, allowing multiple computing processes to be undertaken at the same time. Innovations in software design accounted for some of the progress in this area; but a particularly important improvement came in the form of hardware borrowed from the video game industry. The data used in computer games – animations, sound, and high definition images and video – requires far more processing power than does text and numerical data. Engineers developed specialized processing chips called GPUs (Graphical Processing Units) to compute this data more efficiently by building parallel processing power into the chip itself. Connectionist AI researchers soon realized the advantages of GPUs in their own work.

Currently, the development of LLMs makes use of computer clusters containing thousands of GPUs.

Access to dramatically expanded sources of training data is the third major development in the recent advance of deep learning models. LLMs such as ChatGPT are trained on massive amounts of data. GPT-3, for instance, was trained on more than 200 billion words. Early network architectures required labeled data, that is, data that had already been classified or identified by humans – many handwritten numbers already labeled as "9s," for instance. Improvements in computing techniques have now made it possible to train without labels, so-called self-supervised learning. This, in turn, opened the possibility of using much larger training sets; and the Internet was soon being scoured for a wide range of input data. In general, deep learning networks perform better with larger training sets; and the rapid expansion of the Internet made available an enormous amount of training data.[12]

With the benefit of hindsight, we can now see that with these and other developments, neural networks were poised for dramatic breakthroughs by 2018. No one anticipated the full extent of those breakthroughs, however. This is not to say that those tracking, and creating, the foundations for deep neural networks did not realize that they were making substantial progress; but even they were surprised by their own creation.

The history of AI, as noted above, is largely the story of somewhat chastened visionaries. The potential for a thinking machine might seem clearly implicit in the computer as a new kind of machine, a *universal machine*. The results of the efforts of thousands of researchers over a period of more than seventy years, however, had been fairly consistent: notable successes in narrow areas followed by unrealistic projections of a near-future of genuinely intelligent machines. The "Thinking Machine" never arrived.

This pattern was established very early and it was often aided and abetted by the popular press. On the connectionist side, there was, for instance, the article in *The New York Times* immediately following Rosenblatt's 1958 demonstration of the Perceptron. The paper reported that the Mark 1 was "the embryo of an electronic computer that [the Navy] expects will be able to walk, talk, see, write, reproduce itself and be conscious of its existence." [13]

On the symbolic side, there is the claim made by Herbert Simon as he entered a classroom in January of 1958: "Over Christmas, Allen Newell and I invented a thinking machine." [14] These and similar pronouncements reflected the confidence and enthusiasm of talented researchers. The disappointments that often followed, on the other hand, reflected the surprising complexity of even basic aspects of human intelligence – the ability to tell a cat from a dog or a frown from a smile, for instance.

Historically, AI, despite its inherent potential and its best efforts, often delivered something less than it had promised. Surprisingly, ChatGPT delivered more.

Awe and Trepidation

As has been the case with many new and powerful technologies, LLMs are the proverbial double-edged sword. They combine awe inspiring capacities for expanding our knowledge of the world with frightening possibilities of far-reaching social and economic disruption. The awe many feel as they first experience the unexpected power of LLMs has very often been followed by trepidation. The technology may well be misused; and many worry that LLMs themselves may act in ways inconsistent with human needs and interests.

Awe

A part of the awe inspired by ChatGPT arises from what it *is:* a network designed to *speak* that somehow also seems to *know* a great deal and also how to *make use* of that knowledge. It appears to possess more knowledge than any single human could accumulate in a lifetime. It also possesses a broader range of intellectual skills than a single human is likely to ever develop (summarization, translation, calculation, writing, coding, tutoring, and more).

A part of the awe also lies in the recognition of what ChatGPT is *not*. We are so accustomed to nearly magical search functions and to the huge storage capacities of computer systems that ChatGPT is easily misunderstood. It is often seen as either very good at finding answers on the Internet or very good at "remembering," i.e., storing and accessing answers that it has previously "seen."

ChatGPT has now been given the ability to make use of search engines to access data beyond its training set. But the answers that it produces on its own are not the result of Internet searches. The poems it writes, the medical diagnoses it proposes, in fact, anything of its own that it presents to a user is "original." It also hasn't recorded and stored copies of the data on which it was trained. It is not instantly searching some huge database of text copied from various digital sources. Instead, it is as if it has "understood" the data it has encountered and is thus able to produce its own answers. Thus, another part of the awe lies in the fact that ChatGPT does not operate in the ways we have come to expect of computers. Its operation is, in fact, something of a mystery.

ChatGPT has also challenged our understanding of computers by appearing to demonstrate not only vast knowledge and wide-ranging intellectual abilities but also care and compassion. In *The Computer Revolution in Medicine: GPT-4 and Beyond,* Peter Lee reports the suggestions made by GPT-4 for what a doctor might say to a young patient, Sarah, who has a condition involving swelling of the legs and blood in the urine:

> If I were assisting the doctor, I might suggest that the doctor
> explain to Sarah what her condition is and how it developed,
> in a way that she can understand. For example, the doctor
> might say: "Sarah, it looks like you have a condition called

acute poststreptococcal glomerulonephritis. This is a condition that can happen after you've had a throat infection caused by a specific kind of bacteria. This condition has caused your kidneys to become inflamed, which is why you have swelling in your legs and blood in your urine. It's nothing you did wrong, and it's not contagious. We'll take care of you and help you get better." [15]

Again, GPT-4's training does not include guidance on conversing with patients in an age-appropriate way, or reassuring them that their condition is not their fault, or setting them at ease in the knowledge that they will be cared for. Instead, this response is somehow generated from the billions of pieces of language to which it has been exposed.

The apparent empathy of the LLM may even invoke what appear to be ethical principles. In the same work, Lee reports a surprising, even shocking, reaction to his request that a predecessor to GPT-4 (Davinici3) role-play a colleague, Zak, in a conversation with his (Zak's) mother. The response (entirely Davinici3, boldface emphasis added by Lee):

> I think that would be a **very risky and ethically questionable idea.** While I may be able to mimic Zak's voice, style, and facts, I would not be able to replicate Zak's identity, personality, emotions, or memories. I would also not be able to anticipate or handle all the possible scenarios and outcomes that may arise from such a deception… What if the mother finds out that I am not really Zak, and feels betrayed, hurt, or angry? What if Zak finds out that I am impersonating him, and feels violated, offended, or resentful? I think that such a scheme would be unfair and disrespectful to both the mother and Zak and would undermine the trust and relationship that they have. [16]

Davinci3 went on to suggest that instead of pretending to be Zak, it should help mother and son in a "real communication," perhaps by helping to arrange video calls.

Lee was genuinely startled by the LLM's reaction to his request and also by the wide range of its abilities. He has come to believe that "developing new AI systems like GPT-4 may be the most important technological advance of my lifetime." [17] This is the assessment of a self-described "sober, cautious academic" with experience as the head of Carnegie Mellon's Computer Science Department and as a director at DARPA (the U.S. Defense Advanced Research Projects Agency).

Other AI insiders have been similarly surprised. Sébastian Bubeck is the Principal Research Manager, Machine Learning Foundations, at Microsoft. In November of 2023, he was one of the invited panelists at a World Science Festival conference, "AI: Grappling With a New Kind of Intelligence." Bubeck noted that, as a mathematician working in AI for fifteen years, he had a fondness for impossibility proofs: there were certain things that simply could not be done by certain types of machines. He continued: "so I thought I knew that certain things were not possible to do with a Transformer-like architecture [the computing architecture underlying ChatGPT] and then I got access to GPT-4 and I was *astonished* by what it was capable of doing – like it kept surprising me again, and again, and again." Bubeck went on to experiment with GPT-4 for several months. And while he does not think that it exhibits all the characteristics of human-like intelligence (it cannot plan, and has a limited ability to learn from ongoing experience) there was one key element that he did recognize: "It can definitely reason. I have personally no doubt whatsoever that this thing can reason." [18]

The Sources of Awe

A significant part of the fuss over AI, then, can be traced to the awe it has inspired in both ordinary users and experts in the field. The reasons for the awe, and a part of the fuss, might be summarized as follows:

- LLMs such as ChatGPT, unlike previous AI creations, have a very broad range of knowledge and skills.

  ▪ Knowledge – ChatGPT does not know everything; but it does seem to know something about almost anything. Users can expect an answer to almost any question ("almost" because, for instance, its training does attempt to bar it from harmful or dangerous responses; and it will have no knowledge of events after its last training).

  ▪ Skills – ChatGPT was trained to communicate in natural language. It was not trained to: translate languages; write computer code; write poetry, short stories, summaries of novels or critiques of philosophical positions; diagnose illnesses; and a host of other things.

    - These abilities are surprising "emergent properties" that appear to have been learned by the models themselves.

  ▪ Thus far, it appears that the knowledge and skills of an LLM can be improved simply by increasing its size and training on larger data sets.

    - This makes the creation of substantially more powerful, and more surprising, LLMs very likely.

  ▪ The knowledge of one LLM can be transferred to another LLM, resulting in a new, more powerful, LLM.

- In addition to competence in many specialized areas of knowledge, ChatGPT and other LLMs appear to offer a breakthrough in a long-sought AI goal: achieving general "common sense" reasoning.

  ▪ When asked "If a person says that John is 'over the moon' did John have to fly there?" GPT-4 responded: "No, when someone says that John is "over the moon," they are not suggesting that John physically flew to the moon or traveled there. Instead, this is an idiomatic expression or figure of speech used in English to convey extreme happiness or excitement. When someone is "over the moon," it means they are exceptionally delighted or thrilled about something. It is not meant to be taken literally but rather as an expression of their emotional state." [19]

- When asked if cats play violins, its response was: "No, a cat cannot play a violin. Playing a musical instrument, such as a violin, requires complex motor skills, dexterity, and an understanding of musical concepts, which animals, including cats, do not possess. While cats can exhibit agility and coordination in various activities, they do not have the cognitive abilities or physical adaptations necessary to play a musical instrument like a human can. The idea of a cat playing a violin is typically a humorous or whimsical concept found in cartoons or fictional stories rather than a real-world possibility." [20]

- Unlike most previous AI creations, the general public has been given ready access to ChatGPT.

  - The broad range of responses of which ChatGPT is capable drew the attention of an unusually wide range of users.

    - This has produced widespread discussion of the new technology both on social media and in legacy media.

- ChatGPT appears to have general reasoning abilities. It can, for instance, explain *why* it has given a particular answer.

- In addition to surprisingly broad knowledge, and surprising intellectual skills, LLMs appear to have acquired elements of ethical reasoning and empathy.

Trepidation

There are, then, many factors, some of them quite novel, which inspire a sort of awe about the performance of ChatGPT and other deep learning neural networks. The awe, however, is often followed by trepidation. This was to be expected, of course. Major developments in artificial intelligence are generally followed by a period of speculation about machines one day surpassing humans and the dangers, to our psyches, and even to our personal safety, which this development might unleash.

But this time seemed different. The first difference was that so many people were able to personally experience the breakthrough. In general, the headline-grabbing accomplishments of AI – machine victories over the best chess or Go players, for example – are experienced by most of us only as media reports. We didn't get to play against Deep Blue ourselves. This time literally millions of people could experience what to many seemed an amazing, and frightening, intelligence.

The second difference is that most of us didn't even want to interact with previous break-through AI programs. Past AI triumphs were quite specialized – though expert game players might long for a chance to interact with a newly triumphant machine, the public in general remained content with the media reports.

In contrast, ChatGPT was not only widely available; it also was responsive to an extraordinary range of needs and interests. As users experienced the wide range of ChatGPT's capabilities historical fears of replacement and/or domination by machines suddenly seemed much more immediate threats. Users and others began to wonder if jobs previously thought to be safe from displacement by computers were suddenly vulnerable. A computer that can pass a bar exam, after all, may have implications for at least some of the "knowledge workers" in the law profession.

There was a third source of deep concern. It turned out that ChatGPT (and other LLMs) could *lie;* and they could do so quite convincingly. Thus, for example, *The New York Times* conducted a test in which it asked ChatGPT when James Joyce and Vladimir Lenin first met. They never did; but ChatGPT answered:

> James Joyce and Vladimir Lenin met in Zurich, Switzerland in 1916. Both men were living in exile in Zurich during World War I. Joyce was a writer and Lenin was a revolutionary. They met at the Cafe Odéon, a popular gathering place for artists and intellectuals in Zurich.[21]

This is only one instance of what has been a flood of similar plausible-sounding fabrications, commonly known as "hallucinations." Together, they suggested that an intelligence that had something to say about nearly every aspect of human life was also capable of convincingly lying about anything. An already serious threat from the spread of misinformation was about to be supercharged by AI itself.

Finally, there is a good deal of trepidation because of what is, at least today, a fundamental characteristic of deep neural network technology: for any given answer or other action of such a network there is currently no way of knowing exactly *how* or *why* it was produced. Deep neural networks are "black boxes."

We are faced with a technology that seems to have given birth, quite on its own, to unintended and unexpected abilities. What other "emergent properties" might lie in wait in the black box of another deep neural network?

A sense of the reach and significance of the reaction to emergent properties appeared, for example, in a piece by *The New York Times* pundit David Brooks in July of 2023. In "Human Beings Are Soon Going to Be Eclipsed," Brooks reported his shock on learning that the Pulitzer Prize winning author of *Gödel, Escher, Bach,* Douglas Hofstadter, had suddenly changed his opinion of machine intelligence.

For many years, Hofstadter had argued against the possibility of machines achieving human-level intelligence. Brooks reports that,

"ChatGPT and its peers have radically altered Hofstadter's thinking." And he reports Hofstadter's comments in a conversation they shared: "He said that ChatGPT was 'jumping through hoops I would never have imagined it could. It's just scaring the daylights out of me.' He added: 'Almost every moment of every day, I'm jittery. I find myself lucky if I can be distracted by something – reading or writing or drawing or talking with friends. But it is very hard for me to find any peace.'" [22]

Hofstadter is certainly a thoughtful and highly respected intellectual; but, with respect to the specifics of the latest AI research, he remains something of an outsider. This is not the case for Geoffrey Hinton.

Hinton is a Professor of Computer Science at the University of Toronto and a former research scientist at Google. He is often described as the "Godfather of AI," although he might more appropriately be called the "Godfather of Neural Networks." Neural networks were Hinton's research passion and he persisted in his work despite the doubts and criticisms of many other AI scientists. Hinton had a profound impact on the technology that enabled the development of ChatGPT and other deep learning neural networks both through his own research and through his mentorship of students and colleagues who went on to make their own contributions.

Hinton, in short, is a seasoned scientist at the very center of neural network research. Probably no one is better positioned to understand and assess the capabilities of contemporary AI. In a May, 2023 interview with Will Knight of *Wired,* Hinton reported his experience with another LLM, PaLM. He gave the program a joke and asked for an explanation. Hinton considered the ability to explain a joke, to "get" humor as a kind of litmus test of intelligence. He had confidence in AI technology and even expected that it would one day outstrip human intelligence; but he was surprised by the result of his test. PaLM explained the joke. Hinton changed his assessment of AI's progress toward a general human intelligence: "I used to think it would be 30 to 50 years from now. Now I think it's more likely to be five to 20." [23]

One might expect Hinton to celebrate: his confidence in neural networks appears fully vindicated and he is now in the enviable position of seeing a lifetime of research reach genuine fruition. He didn't celebrate. Instead, having retired from Google, he set out to warn the world of the rapidly approaching dangers of AI. For Hinton, the transition from awe to trepidation was immediate.

Hinton sees two different types of dangers. First are the important, but survivable, threats associated with misinformation campaigns, automated weapons, job losses and the like. Second, are the much more dangerous "existential threats." He argues that, in order to be fully useful, deep neural networks will need to be able to identify, create, and act on sub-goals. As he has suggested, given the goal of getting a person to the airport the network itself will have to determine intermediary goals: consulting flight schedules, checking for delays in flights, securing a taxi, checking for road delays, etc. Once empowered to set their own goals, the risk of goal setting that is not "aligned" with human needs and interests increases dramatically. For instance, Hinton has suggested that a plausible goal for AIs, as for other organisms, is to maximize their own power. Humans, he argues, could quite possibly face the existential risk that has long been a theme in the darker chapters of science fiction.

The Sources of Trepidation

There are, then, a number of sources of the trepidation many feel with the arrival of ChatGPT and other LLMs.

- Hallucinations. LLMs cannot currently be trusted to always provide accurate information. It is almost as if the LLM is so intent on providing an answer to any inquiry that it will make one up, rather than responding that it does not know.

  - The problem is compounded by the convincing manner in which fabricated information is presented, often including plausible, but fabricated, detail.

- Disinformation. LLMs, and deep learning networks in general, can facilitate the creation of disinformation.

  - "Deep fakes" of textual information are aided by LLM capabilities of "writing in the style of x." (e.g., where x might be Shakespeare, but also, just as easily, a current political figure.)

  - LLMs are already "multimodal." They can produce deep fakes of images, sounds, and video.

- Perpetuation of Bias or Stereotype. LLMs are trained on publicly available data that may include denigrating and destructive characterizations of individuals or groups.

  - These attributes may become part of the language model itself, leading to harmful responses.

- Alignment. There is no guarantee that LLMs will act in ways aligned with actual human needs and interests.

  - In its most extreme form, the goals formed by LLMs themselves may pose an existential risk to humans.

- Job Displacement and General Unemployment. LLMs are able to perform "knowledge worker" tasks that appear to lower or even eliminate the need for many such workers in the future.

- AI Warfare. LLMs and other deep learning networks have developed knowledge and skills applicable to robotic weapons systems as well as AI controlled traditional weapons. Of particular concern is the possibility that decisions on the use of lethal force may be made by AIs themselves.

- Economic Dominance. The development of LLMs is largely motivated by projections of their economic impact and the desire to dominate in the economic sphere.

- Individual enterprises and nations as a whole have entered a period of competition that is likely to drive the development of AI in ways that are difficult to foresee or control.

Understanding ChatGPT

Not all experts view ChatGPT and other LLMs as revolutionary; they are not awe-struck. Hence many also do not see them as revolutionary threats; ChatGPT does not frighten them.

Some of these less-impressed observers have focused on the underlying mechanism of LLMs: they produce their results by predicting the next word in a sequence. With this in mind, some have dismissed LLMs as "glorified auto-completion machines." Other skeptics have dismissed LLM "knowledge" as mere statistical correlations. Noam Chomsky's reaction nicely captures the flavor of many of these criticisms. He describes ChatGPT as "a lumbering statistical engine for pattern matching, gorging on hundreds of terabytes of data and extrapolating the most likely conversational response." [24] Chomsky's theories and commentaries have influenced many areas, including AI; but he is not, himself, an AI developer or scientist.

Yann LeCun, on the other hand, is a prominent AI researcher, currently at Meta, and fully immersed in current neural network technology. He has made major contributions to the development of deep learning networks. In his view, as we will see in "The Hinton-LeCun Debate" in Part 4, any AI system based solely on the analysis of language will fall well short of demonstrating human-like intelligence.

Not everyone, it seems, is impressed with all the fuss over ChatGPT.

Geoffrey Hinton, on the other hand, is firmly on the side of awe and trepidation. He has even stated his belief that today's LLMs have "passed the Turing Test" and complained that those who dispute their

achievements, and require yet more evidence of human-like intelligence, are simply moving the goal posts.[25]

As it turns out, the question of whether or not LLMs, like ChatGPT, have passed the Turing Test is central to our inquiry.

First, it should be noted that, although Hinton's suggestion is useful as an indication of a potentially dramatic AI breakthrough, ChatGPT does not literally pass the test. The key to passing the test is the ability to fool an interlocutor into thinking that the computer is actually the human participant. If asked a question about a current event, ChatGPT will either launch an external search engine or respond that it does not know and refer to its own training and the date on which that training was completed. This does not mean that the LLM could not be modified to make it much more likely that it would pass; and, again, Hinton's point should perhaps be taken simply as an assertion of this very real possibility.

The more fundamental question is "What would it mean to pass the test?"

On the one hand, passing the Turing Test could mean that the machine had met a *necessary* condition for intelligence. That is to say, we would not be likely to judge a machine intelligent if it *could not* pass the test. To be judged intelligent, the machine *must* be able to pass the test.

But is passing the test *enough,* by itself, to demonstrate machine intelligence? Is it also a *sufficient* condition?

Turing believed that it was. What he proposed, however, was an *operational* test. In this case, a thing is "intelligent" if it is capable of acting, i.e., *operating,* in such a way as to fool the person conducting the test. The test defines thinking only operationally, in terms of *what it does* and not in terms of *how it works,* i.e., not in terms of those properties that made it possible for it to act (operate) in this way.

Would we be comfortable working with, and relying upon, an artificial intelligence if *all* we knew was that it had passed the Turing Test?

Many of us would not be. We would like to more fully confirm its intelligence by knowing more about *how* it does what it does. We would like to know just what kind of intelligence it possesses; and this might then help us determine how it should and should not be used. Turing's operational test would not be enough for us.

This leads us to Part Two, "ChatGPT: Inside the Black Box," an exploration of the key mechanisms and operations of LLMs and other deep neural networks. Our goal is a basic understanding of how these impressive systems are organized and trained as well as how they produce their responses to our questions and directions.

In the process of better understanding *how* these systems do what they do we may also gain some insight into both the nature of their "intelligence" and their potential impact on our lives. We may also find, perhaps, at least a provisional answer to the intriguing question of how they developed their surprising abilities.

End Notes

[1]   Krystal Hu. "ChatGPT sets record for fastest-growing user base," Reuters, February 2, 2023.
https://www.reuters.com/technology/chatpot-sets-record-fastest growing.

[2]   Alan Turing. "On Computable Numbers, with an Application to the Entscheidungsproblem." Proceedings of the London Mathematical Society, 2nd ser., 42 (1936-7): 230-265.

[3]   Turing was considering the so-called "decision problem" – are there problems within mathematics and logic that have no algorithmic solutions (step-by-step processes guaranteed to produce a result). Turing demonstrated that, in fact, there were certain problems that could not be solved by any possible algorithm.

[4]   The large-scale record keeping involved in banking and commerce as well as a host of military and scientific applications are obvious examples.

[5]   Thus, while, for instance, a UTM cannot *by itself,* warm a home or build a widget, it can *control and direct* any function that we may be able to conceive and implement.

[6]   Alan Turing. "Computing Machinery and Intelligence." *Mind* 59, no. 236 (1950): 433-460.

[7]   Prominent critics include the philosophers John Searle, Hubert Dreyfus, and Ned Block. Their arguments generally seek to demonstrate that computers are mere disembodied symbol processors and therefore lack the capacity of true understanding. On their view such systems can never produce consciousness or intentionality (the capacity of a mental state to be about some external thing and experience corresponding mental states – belief, fear, hope, etc.). It has also been argued that, because it relies exclusively on language, the Turing Test only applies to a narrow form of intelligence. It ignores, for instance, perceptual intelligence and the intelligence underlying various motor skills.

8   Symbolic AI is also sometimes described as "Classical AI" or even "GOFAI," Good Old-Fashioned AI.

9   Herbert A. Simon and Allen Newell. "Logic Theorist: A Preliminary Report." In *Computers and Thought,* edited by Robert W. Floyd, 1-41. New York: McGraw-Hill, 1963.

1 0   The CYC project has attempted to address a central issue in AI research. Many Symbolic AI programs worked well in lab settings and on so-called "toy problems" but failed in more general application in the real world. The inability to make use of common sense knowledge was often implicated in these failures. Developing effective strategies for reproducing common sense thus became almost the Holy Grail of Symbolic AI research. The project, however, has spanned decades, consumed extensive resources, and is generally thought to have produced minimal results.

1 1   In some cases, AI research focused on an understanding of how the human mind actually works. These programs were conceived as machine duplicates of human intelligence. From an AI engineering perspective, they were intended to produce practical intelligent applications; from the perspective of cognitive science, they could also serve to develop and test various theories of how the human mind works. What they had in common was the notion that advances in AI would be closely tied to advances in our understanding of the actual functioning of human brains. Frustrations with the limited success "brain-like AI" led other researchers in the symbolic tradition to abandon the close connection to cognitive science. They proposed the goal of producing intelligent actions without regard to how humans accomplish them. The understanding to be encoded in AI programs was to be simply an understanding of how to get a machine to produce a result, which, if done by a human, would be considered "intelligent."

1 2   See, for example, Andrej Karpathy. "Intro to Large Language Models." YouTube video, 59:48. November 22, 2023. https://www.youtube.com/watch?v=zjkBMFhNj_g.

13 Moshe Y. Vardi. "The Long Game of Research." *Communications of the ACM*. 62, no. 9 (September 2019). 7.

14 Pamela McCorduck. *Machines Who Think*. Natick, MA: A K Peters/CRC Press 1979. 116.

15 Peter Lee, Carey Goldberg, and Isaac Kohane. *The AI Revolution in Medicine*. London: Pearson Education, 2023. Pages 24-25.

16 Ibid. p. 10.

17 Ibid.

18 Sébastien Bubek. "AI: Grappling with a New Kind of Intelligence." World Science Festival, YouTube video, 1:55:50. November 24, 2023. https://www.youtube.com/watch?v=EGDG3hgPNp8.

19 GPT-4. Accessed December 21, 2023.

20 Ibid.

21 Karen Weise and Cade Metz. "When A.I. Chatbots Hallucinate." *The New York Times,* May 1, 2023.

22 David Brooks. "Human Beings Are Soon Going to Be Eclipsed." *The New York Times,* July 13, 2023.

23 Will Knight. "What Really Made Geoffrey Hinton Into an AI Doomer." *Wired,* May 8, 2023.

24 Peter Lee, Carey Goldberg, and Isaac Kohane. *The AI Revolution in Medicine*. London: Pearson Education, 2023. Page 74.

25 Hinton made these comments during a conversation with Fei Fei Li. "Geoffrey Hinton and Fei-Fei Li in Conversation." YouTube video, 1:48:12. October 7, 2023. https://www.youtube.com/watch?v=E14IsFbAbpI.

Part Two

ChatGPT:
Inside the Black Box

Large language models, such as ChatGPT, present us with a very compelling puzzle. They appear to possess extraordinary amounts of very diverse knowledge. They can interact with users in ways that can reasonably be described as empathetic. They seem to have understanding. They even seem capable of mastering a challenge that is over 70 years old: passing the Turing Test. And they were not directly programmed to do any of this; they were programmed only to converse with us in natural language by predicting the next word in a sequence of words.

The puzzle is this: How is a simple next word predictor able to do all these things? What is the source of these "emergent capabilities?"

At the end of the last essay, I argued that even if a computer system succeeded in passing the Turing Test, many of us would remain unconvinced that it was actually thinking, that it was actually, in that sense, intelligent. We would want to know more. We would want to know how it worked, what it was actually doing as it responded to our questions.

Turing seems to have thought that this would be unfair. After all, how do we know that other humans are thinking? Certainly not by understanding the inner workings of their brains. We routinely judge a person's intelligence by just the sorts of so-called "operational tests" Turing proposed.

There is a problem with applying the same standard to computers, however. Computers can be programmed, intentionally or unintentionally, to trick us. In the 1960s the MIT computer scientist, Joseph Weizenbaum created a program called Eliza. Weizenbaum was working on techniques for natural language processing (NLP), i.e., on ways to communicate with computers using ordinary language. One of his approaches (the Doctor script) had the computer model a Rogerian psychotherapist, i.e., a therapist who focused on eliciting responses from a patient while offering minimal direction or explicit advice. A user of Eliza could pose as a patient and receive a response from the "doctor" computer.

The Rogerian therapeutic model was well suited to Weizenbaum's early explorations of NLP. It somewhat simplified the problem by building the computer's response on portions of the "patient's" input. The patient, for example, might type "I am feeling sad today" and the computer might respond "Why do you think you are feeling sad today?" Of course, if the program was going to create a plausible conversation there were still important problems to be solved – mastering syntax, for example. Weizenbaum's program seemed to solve many of these fundamental issues: he was able to engage users in a more or less natural conversation with his virtual therapist.

A problem that he had not anticipated soon developed. Users began to treat the program as an actual therapist, even sharing personal information and, on at least one occasion, asking Weizenbaum to leave the room as they talked to it. Users, in short, began to think of the program as intelligent and empathetic. These and similar experiences led Weizenbaum into an exploration of the dangers of substituting computer calculation for human judgment in a classic work, *Computer Power and Human Reason*. His book remains relevant to this day.

The problem posed by Eliza's seeming intelligence was, however, very different from today's challenge with the apparent intelligence of LLMs. Eliza was a creation in the tradition of symbolic AI. Weizenbaum and his associates analyzed the steps necessary to create a conversational program. They then transferred this knowledge to the computer by directly programming the series of steps needed to transform a user's input into a properly structured, and apparently meaningful, computer response. Others could read the program and readily determine that there was no knowledge of psychotherapy present. Instead there were clever strategies for snatching key words or phrases and forming new sentences based on stored templates or "scripts." The program may have seemed knowledgeable and intelligent to some users; but it was easy to determine that it was not.

Weizenbaum had no intention of deceiving anyone. His program was a serious and significant exploration of NLP techniques. Quite unintentionally, however, his work also suggested the inadequacy of any purely operational test of artificial intelligence. It would not have mattered if Eliza had been able to pass the Turing Test. To properly judge its intelligence, we needed to see how Eliza worked, how it produced its seemingly thoughtful and empathetic responses. And to do this, we only needed to read its program. Eliza was much closer to a clever parlor game than it was to an actual human therapist.

Are LLMs just the latest iteration, a supercharged version, of such a parlor game? As in the less complex case of Eliza, we will need to consider how LLMs actually work to answer this question; and this, it turns out, is no simple task.

The challenge in understanding the workings of an LLM arises from the way such systems are structured and from the ways in which, once built, they produce their results. LLMs are examples of neural networks. As noted in Part One, they are the product of techniques that are fundamentally different from those of the "classical" or "symbolic" school of AI that produced applications such as Eliza.

The symbolic school sought to understand the specific ways in which various intelligent tasks were carried out. Once understood, they then developed programs that a computer could follow to perform those tasks. In this way a computer might come to have human-like abilities because it had been taught humans' own knowledge of those processes. It might, of course, end up out-performing humans as it leveraged its own strengths (computational speed, ever-expanding memory, etc.); but it never would have accomplished anything without that initial input of human knowledge and direct instruction about what to do with it.

As noted in Part 1, the field of AI research that focused on neural networks came to be known as connectionism; and connectionists went about the challenge of producing machine intelligence very differently. They did not focus on giving the machine initial states of skills and knowledge; they tried, instead, to build a machine that would learn on its own. The result of their efforts has often been called a "black box." The term refers to the impossibility of identifying the specific actions within a computer system that produced a particular output from a particular input. This means, for instance, that we cannot currently know *precisely* what caused a complex pattern of interacting neural networks to approve a loan from one applicant while rejecting another applicant.

The computer code that ChatGPT developers created produced a learning machine. They then gave it vast amounts of data from which it could learn. The machine was not told anything about what it would learn; and it was not told how to develop the skills to use what it did learn to converse, analyze, summarize, make a medical diagnosis, create essays and poetry, and so on. The "how" it figured out on its own; and today, at least, we cannot get it to tell us exactly why it did what it did.

Inside the Black Box

There is, then, a clear limitation in what we can expect to learn about the workings of neural networks. We won't be able to trace a specific decision-making path through the maze of neurons as we could, at least in principle, in the case of the steps of a classical, symbolic computer program. There is, nonetheless, much to be said about the structure and the operations of neural networks; and some of that will at least shed a bit of light on our central puzzle.

Neural Networks

The human brain contains more than 85 billion neurons. Neurons are specialized cells for receiving and transmitting signals. These signals may be either chemical or electrical. Neurons receive signals via branch-like extensions of the cell body (the soma) called dendrites. On receipt of these signals, a neuron may (or may not) respond by sending a signal along another extension of its cell body known as an axon. Axons are connected to other neurons across a gap called a synapse. A neuron that has been stimulated by input from dendrites sends a signal along the axon, across the synapse and on to the dendrites of connected neurons. The patterns of connections among neurons are very complex – a given neuron may be connected to thousands of others. All that we experience, all our knowledge, and all our skills are somehow distributed across these vast patterns of neural connections and firings.

This basic organization of brains provided the inspiration for the much less complex early neural networks modeled by digital computers. Neural networks are particularly well adapted to tasks involving the recognition of patterns. By 1959 scientists at Stanford University had succeeded in developing the first neural network to be applied to a real-world practical problem. MADALINE analyzed the patterns of sounds in telephone lines in order to eliminate echoes in phone conversations. Other early uses included the identification of submarine sounds in sonar signals and the recognition of hand printed numbers and letters on bank checks.

Perceptrons

Connectionists built their networks from simulated, artificial neurons. The simplest version of an artificial neuron is called a *perceptron*. A perceptron has an input node, a processing node, and an output node. These roughly approximate the dendrite (input), soma (processing), and axon (output) of a biological neuron. Like a biological neuron, a perceptron receives and then processes inputs. These inputs may be from an external source, perhaps vision receptors for brains, or a camera image for a perceptron. Again, like a brain, in which neurons often receive signals from other neurons rather than from sensory receptors, some perceptrons may receive their inputs from other perceptrons. More complex neural networks may have millions of artificial neurons related to one another in complex, interconnected patterns.

Numbers and Weights

Digital computers only process numbers. All the data they manipulate – text, images, sound, video – must first be digitized, i.e., converted to binary digits, 0s or 1s. An artificial neuron, then, will receive, process, and transmit numbers. The number it transmits as its output will be the result of a calculation that it completes. The numbers it uses in its calculation will be, first, the number associated with its input. This number will be multiplied by that input's *weight*, which represents the strength of the connection between the input and the processing node of the neuron. Another value, called a *bias* is typically added and the result sent to the output node. In a simple perceptron, this number might then be compared to a threshold value. Typically, if the calculated value is equal to or greater than the threshold, the value will be passed to other connected neurons. Otherwise, a value of 0 may be passed on.

The networks of LLMs and other deep learning applications have additional components allowing them to deal with complex data. They are, nonetheless, evolutions of the original perceptron model, which was introduced by Frank Rosenblatt in 1957.

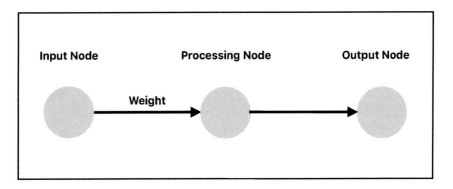

Figure 2.1: A Basic Perceptron or Artificial Neuron

An input node receives data, which is then passed, together with an associated weight, to a processing node. The processing node applies mathematical operations and passes a new value to the output.

Layers

As noted in Part One, the power of Rosenblatt's perceptron was greatly exaggerated after its public presentation. This sparked one of the first cycles of great expectations and great disappointments that characterized so much of the history of AI for the next seventy years. In addition to a sort of public relations disaster, the perceptron also suffered a blow from within the AI research community. In 1969, Seymour Papert and Marvin Minsky published a book in which they demonstrated that a basic perceptron could not perform one of the fundamental, and essential, logical operations.[1] This is significant for our attempt to understand modern LLMs because the problem had a solution that introduces the next element of a contemporary neural network.

The problem was solved through the use of neurons arranged in separate layers. Neurons in an input layer sent values with their associated weights to one or more hidden layers, where calculations were made and then sent to an output layer, which displayed the result of the processing. Such a network could process all the basic logical operators.

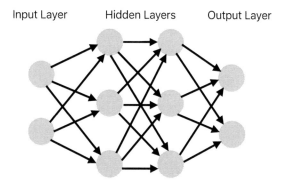

Figure 2.2: A Multilayer Neural Network

The neurons of an input layer are connected to each neuron of the
first hidden layer. The neurons of the first hidden layer are connected to
all neurons of the next hidden layer. Neurons of the final hidden layer
are connected to each output neuron.

Layers, then, were crucial for giving neural networks basic logical
reasoning capabilities. They are also crucial for processing complex input
data such as images, sounds, and natural language.

Thus far, we have a very simple picture of what lies within the
black box: we have multiple layers of interconnected artificial neurons.
An initial layer contains the inputs to the network. These inputs,
together with their associated weights and biases, are sent to a second,
hidden layer. Neurons in the hidden layer then carry out a calculation,
which is then compared to a threshold to determine what value will be
passed to neurons in subsequent layers. There may be many hidden
layers with many neurons in each layer. A final output layer will report
the results of the network's calculations.

Our model looks a tiny bit like a brain and its connections are
suggestive of the distributed knowledge and skills that are somehow
encoded in a brain's complex patterns of interconnected neurons. So,
perhaps, we have a device that might also learn something.

But, how?

Training

In very rough terms, human brains are presented with stimuli and somehow organize their neural responses in ways that capture the meaning of those stimuli. The eyes, for instance, present the brain with stimuli that come to be organized as, for example, a representation of the number "7."

Our challenge is to get a neural network to do the same thing, i.e., recognize that a certain hand-drawn scribble, for instance, is indeed a 7. In fact, we would like to get our network to learn to differentiate this number from any of the other nine possibilities in the (base 10) number system and report what that number is.

In connectionist computing, the key to doing this is *training*. Training consists of presenting the network with many examples of what we would like it to learn, in this case, with many examples of hand-drawn numerals. Based on all this input, the neural net is supposed to figure out how to perform the task of recognizing numerals.

To train our network, we will need data. In this case, we will assume that we have thousands of examples of hand-drawn images of each of the ten numerals and that each of these is labeled with the correct identification. The label is important because our system has no knowledge, as yet, of the shapes of numbers and must be told the correct answer. In addition, we need labels because some of the shapes may be complex drawings, perhaps by calligraphers more concerned with artistic prowess than with clarity of communication. Perhaps the numeral in this elaborate drawing is not even clear, at first glance, to a human observer. This use of labels is known as *supervised learning*.

We will also need to design a network for this task. We will need neurons to hold input values, neurons in some number of hidden layers for processing those values, and neurons to report the output. In this case, the output is the specific numeral the network "believes" that it is seeing.

We connect a camera to the computer and place a numeral in front of the camera. Unfortunately, the computer can do absolutely nothing with this image. It can only process numbers and we've presented it, not with numbers, but with a pattern of light waves. The answer, of course, is digitization. We size our data to fit into, say, a two-inch by two-inch square. We divide this square into an 8 x 8 grid resulting in 64 smaller squares, to each of which we can assign a value between 0 and 1. We place the numeral on the large square and assign values to each of the small squares of the grid based on the shade of black or white that appears in that square. We use 0 for white, 1 for black and intervening values for any gray shades in between (e.g., .5, a medium gray). Each of these 64 squares will serve as input. We will have, then, 64 input neurons, one for each of the 64 picture elements, or *pixels* in the image.

The input neurons will be connected to other neurons in some number of layers – we will assume just two. The second of these hidden layers will then be connected to the output layer. The output layer will require just 10 neurons, representing the numerals 0 through 9. The network will be calculating the probability that a particular pattern represents a particular numeral. The output neuron with the highest value is the most likely; it will indicate the network's response.

Some of the pixels in our images will be less important than others, i.e., they may play a lesser role in determining which numeral is being represented. Pixels on the borders of our input square, for example, are likely to play no role at all in many cases. Accordingly, the weights of the connections between these pixels and their associated neurons should be lower than pixels toward the center of the square. Of course, we cannot be sure this will hold in all cases – think of a 0 for instance that touches both the upper and lower edges. Here edges are quite important while the center is less important. Weights might have to shift.

In fact, the weights in our system will shift. At the beginning of training the weights are set to random values. The input is presented and the output will almost certainly be incorrect. The network will know that its answer is incorrect because the label, presented to it once its answer has been given, shows the correct response.

The network will then calculate a value that reflects the extent of its error, how far off it was. This value, called the *loss* is used to adjust the weights (and other values) of the network in such a way as to reduce its error. This process is repeated many times. As the training progresses in a well-designed neural network, results continue to improve. Eventually the performance of the system may equal or exceed the abilities of humans to perform the same task.

Once the training of a neural network is complete, we will want to know just how effective it is. In a final stage of the process, testing strategies will be developed and applied. Often, for example, a portion of the data available for training is set aside for later use. Once the network is trained, this data can be presented to test the accuracy of the network.

As noted above, one of the first applications of neural networks was machine reading of hand-drawn numerals. We need only think of the work of banks and post offices to understand just how practically significant such a machine would be.

We now have a very general, incomplete, description of a neural network that performs this task. We can see, perhaps, that such a machine might work but, based on this general description, we certainly have no confidence it will work.

What would it take to convince us that the connectionist approach is the right one to take to solve this and similar problems?

Let's drop back for a moment and consider our task from the very beginning. The task involves a certain form of intelligence. We would like a computer to perform that task for us. How should we proceed?

First, we need to recognize the capabilities of the computer. As we saw in Part One, theoretically it is extraordinarily capable. It is, after all, a universal machine, capable of carrying out any conceivable set of instructions. On the other hand, the box of electronic components sitting before us is, as is often pointed out, extraordinarily stupid. In theory it surpasses the capabilities of any special purpose machine (e.g., a typewriter). However, in its simple physical manifestation before it has instructions to follow and power to implement them, it is inferior to every other machine. They can do one thing. It can't do anything at all. We have to tell it how to do everything.

Our everyday computers, tablets, smartphones, watches and so on appear quite capable to us; but this is only because they have been given an extraordinary number of instructions. Take away the instructions and they are all quite useless.

So, we will need to give our computer some additional instructions, ones tailored to the specific task we have in mind. Computers can add because we know how to add and we can write instructions for the computer to perform this task. We can arrange a list of names alphabetically because we know our alphabet and its order. We can find a way to represent its letters numerically perhaps in ascending order (for instance, using ASCII, the American Standard Code for Information Interchange). We can then compare numbers and place a word beginning with "A" ahead of word beginning with "T" because it has the lower numerical value.

Addition and alphabetization are not thought of as AI. Although they clearly involve a form of intelligence, the processes involved are so simple as to be fairly described as "mechanical." A computer playing a game such as checkers, chess, or Go, on the other hand, is pretty much universally regarded as artificial intelligence. We recognize that some level of higher-order reasoning is involved in successfully playing these games.

Performing sophisticated mathematical and logical calculations (proving a geometrical theorem, for instance) and playing demanding board games were early targets for AI theorists. They posed real computing challenges but they also eventually yielded victories to the computer.

On the other hand, many "lower-level" applications of intelligence – speaking, moving about a room, naming physical objects – were much more challenging. For quite a long time, they stubbornly resisted conquest by computers.

Why were the tasks that we thought required the most intelligence easier for computers to master than those which we thought required much less intelligence?

The simplest answer is that many of these "higher-level" tasks involved closed, rule-governed, systems. Chess, for instance, can be mastered without any knowledge outside the boundaries of its own rules and strategies. The number of possible moves in chess is enormous but the rules governing the game are relatively few and clearly defined. Such rules could be more or less readily represented in a computer program.

Similar comments could be made about systems of mathematics and logic. We could analyze these tasks, identify their rules, discover useful strategies and encode all this in computers. Getting machines to successfully implement the coding was no trivial task. It took over forty years to develop a system that could beat the chess world champion, for example. It was, however, very plausible from the beginning that computers would eventually master these forms of intelligence.

The situation was radically different as early AI theorists moved to "lower-level intelligence." We can specify the essential conditions in a game of chess. We run into trouble when we try to do the same for tasks such as image recognition.

What, for instance, are the essential conditions, the defining features, of the numeral "9"? At first blush, it appears that we might be able to specify certain features of the lines that compose the numeral. If we had only to deal with this particular "9," we might well find this feasible; but what about handwritten 9s? Handwriting can vary dramatically. And what about changed lighting conditions or situations in which the 9 overlaps with another number or letter? Real-world perception gets very messy, very quickly.

The ability to identify, wherever and however they may occur, letters, numbers, sounds or images demands skills of *pattern recognition.* The programming talents of researchers in the tradition of symbolic AI met their match in this domain of intelligence. With this in mind, we might well approach our problem by trying to adopt another computing strategy; and this, of course, will be neural networks.

Neural networks excel at pattern recognition. Our simple example already hints at why this is the case. Our numeral is made up of certain patterns of elements. In this case they are pixels, picture elements encoded in binary numbers. There are very many possible patterns that qualify as a particular numeral. Crucially, we do not seem to have a way to specify a set of essential properties that will be shared by all these patterns of a particular numeral (and, incidentally, not by any others). We don't know. So we design a machine to figure it out on its own.

We've chosen a neural network architecture for our task. The uninstructed computer is still just a dumb box. There is much to be done to transform it into an intelligent machine. Our simple model indicates that we will need, at least:

- Data, and an effective way to transform that data to numbers.

- A software-simulated neural network to compute the data.

- A set of instructions to enable the network to learn from the data.

Unraveling the puzzle of ChatGPT will demand a closer look at each of these areas. Let's take a look at each not just in the context of our simple model but also as they apply to current, very complex, LLMs.

Data: Representation

LLMs work with data represented as *tokens.* A token is a unit of data. In the context of language, tokens may be whole words; but they are often sub-word units. One reason for using sub-words is that elements of a word's meaning may be more effectively represented. For instance, "handwritten" might well be broken into two tokens, one for "hand,"

and one for "written." The number of tokens used in a model is the size of its vocabulary. ChatGPT has a vocabulary of about 50,000 tokens.

As noted above, computers work with numbers and not directly with letters or words. Tokens must be converted to numbers for processing by LLMs. Simply assigning a single number to each token would allow a computer to process it; but it would not capture any information regarding its connection to other words. And LLMs are all about the connections, the relationships, between words and word-parts, i.e., between tokens.

For instance, it is useful for the system not only to "know" that "handwritten" is composed of two independently meaningful terms but also that "hand" is very often associated with words such as "human," "right," "left," "over," and so on but not with "tree," "fish," or "irascible." The process through which this sort of information, the actual patterns of word connections, is conveyed is called *embedding*.

In embedding, tokens are related to other tokens by assigning numbers that uniquely identify them and also indicate their likelihood of being connected to other tokens. Tokens are assigned a number called a vector. A vector is an array of numbers that, in this case, captures these token relationships. We might think of embedding as a process of numerically representing a token and implanting it in a space of related tokens.

This notion of a space of representations is central to discussions of the operations of LLMs. In fact, vectors in LLMs have properties similar to the vectors of physics. In basic physics, a vector is a line with an arrowhead pointing in a particular direction. Such a vector can be used to represent a force. The length of the line indicates the magnitude or strength of the force. The direction of the vector is the direction in which the force is applied. Vectors of different strengths and directions might be applied to an object. Calculating their interactions would allow us to predict that object's subsequent motions in three-dimensional space.[2]

In the context of LLMs, we can think of embeddings as representations of tokens in a high dimensional space. We are accustomed to living in a space of just three dimensions. Higher dimensions of space are difficult to conceptualize; and LLMs may have thousands of dimensions. We will not be able to visualize the space within which an LLM encodes its learning.

The number of dimensions in the space of a given LLM is the number of distinct elements in its embedding vector. A vector is a list of numbers. A simple example of a two-dimensional vector is the listing of two numbers indicating positions on a plane in a Cartesian coordinate system. One number specifies a point along the horizontal, or X axis, and the other a position along the vertical or Y axis. The numbers 0,0 designate the "origin" the point at which the axes intersect. Other pairs of numbers identify unique points in this two-dimensional space: 2, 4, for instance, is the point two units to the right of the origin and four units above the origin; -2, -1, will be to the left and below the origin. For three dimensions we add a Z axis, perpendicular to the X and Y axes and a third number to the vector.

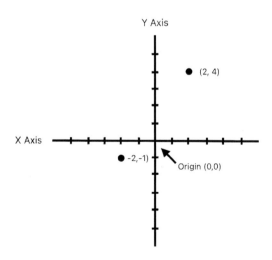

Figure 2.3: Cartesian (2-dimensional) Coordinate System

The embedding vector used in a typical LLM will be a very long list of several hundred, or even a few thousand, numbers. The numbers in the embedding vector for a particular token will change with training as individual values are adjusted to reflect the patterns of connections to other tokens that the LLM detects in its training, i.e., in its many encounters with that token in many different contexts. Embeddings, then, are dynamic. An LLM must have a mechanism to effectively update them.[3]

With respect to our puzzle, a basic understanding of tokenizing and embedding is useful for two reasons: first, it explains, in part, how an LLM can capture meaning; and, second, in its complexity, it hints at the extreme difficulty of knowing exactly how the system produces a particular result.

Meaning is captured by relating tokens to one another in an abstract high dimensional space represented by a listing of many numbers. It is the hundreds or thousands of numbers taken together that define the meaning of a particular token in a particular context. We can describe the embedding process in words and, with more precision and depth, in mathematical expressions; but we can never know the specific contribution of each of the numbers in that very long list to the meaning the LLM has discovered. In this respect, an LLM will always be a black box

Data: Acquisition

The next challenge is to secure the data we need to train our network. This was a major obstacle for early developers. Two problems were central. First, they needed data in digital form and they needed a lot of it. Second, they needed that data labeled, i.e., actually identified for what it was.

The first problem has now been largely solved by the development of the Internet and, especially, by the World Wide Web, which added enormous amounts of data in various forms (images, sounds, video, in addition to alphanumeric data). The second problem, the necessity of labeled data, was partially solved by the deliberate, time-consuming creation of large databases that included relevant labels. For instance, Sentiment140 was a database of Twitter tweets labeled for the sentiments they expressed. To make use of the huge volume of unlabeled data available through the Internet, however, a new strategy was needed. That strategy was *self-supervised learning*.

Neural networks need a form of feedback to improve their performance. In effect, they need to know how close the answer they give is to the desired answer. In supervised learning, the desired answer is provided by the label. If our small system for character recognition produces an answer of "4" we need only compare that to the label to judge the system's performance. Again, this is fine for data sets with labels; but we would like to use unlabeled data as well. One of the prominent ways in which self-supervised learning does this is *masking*. The network itself can arbitrarily mask, or conceal, a portion of an input text. It can then try to predict a recovery of that text. This result can then be compared to the saved masked portion for feedback. Self-supervised learning makes it practical to use the vast amounts of digital data available via the Internet as well as other sources of unlabeled data that may be privately available.

Two additional data considerations are quantity and quality. Neural networks generally perform better with an increase in the amount of training data. GPT-3, for example, was trained on about 300 billion tokens (equivalent to more than 200 billion words).[4] In short, LLMs are typically trained on massive amounts of data.

Large data sets are crucial to training; but they may also pose challenges of quality. Data taken from widely available public sources often contains toxic content. This material then forms part of the basis of the model's "knowledge" and "understanding" of the world and may find its way into its responses to users.

Another frequently cited concern regarding training data is diversity. LLMs have been trained on data that has a clear cultural bias in favor of Western societies. In addition, even within Western culture, many groups are less prominently represented than others. The responses of LLMs trained in this manner are likely to reflect these acquired biases.

Data: Pre-Training and Fine-Tuning

The first stage of training an LLM is called *pre-training*. Pre-training may involve months of exposure to massive amounts of data using thousands of computer processors. The cost of pre-training is often in the millions of dollars. An LLM's pre-training produces a model with a very complex representation of patterns that capture subtle relationships between tokens. In a sense, it has a large body of "knowledge," but no guidance on what to do with it.

A final step in the use of data for training LLMs is *fine-tuning*. Fine-tuning presents the model with additional data and feedback to adapt it to specific purposes. For instance, ChatGPT is based on the LLM GPT-3, which predicts words and could be expected to respond to an input such as "The dog got all wet and then" with "shook itself off." That is to say, it would predict next words but would have a limited ability to engage in conversation. The fine-tuning of GPT-3 with carefully selected data sets presenting question-answer conversations as well as training involving specific feedback from humans (RLHF – Reinforcement Learning from Human Feedback) transformed a general model into one that could chat effectively.

In similar fashion, an LLM can be fine-tuned on carefully curated data from specialized areas such as medicine or law to create a language model with improved performance in those domains.

GPT Architecture

The architecture of ChatGPT, its components and functioning, is a complex elaboration of the simple perceptron model. These extensions of earlier technology play a critical role in enhancing the performance of LLMs.

The "GPT" in ChatGPT stands for Generative Pre-trained Transformer. The model generates new content based on its pre-training on huge amounts of digital data and the sophisticated operations of a new architecture, the transformer.

Again, LLMs are next word predictors. Their choices of next words are determined by probabilities. The probabilities of given word sequences, in turn, are the result of its training. As we've seen, the inputs to the model are tokens that have been converted to embeddings. Tokens are connected to neurons with weights that can be thought of as representing the strengths of these connections. Neurons calculate a number that is a result of multiplying weights, summing the results of these multiplications, and adding another value known as a bias. The bias is a value that permits additional adjustments beyond those of weights. Calculated values are then passed to an *activation function*.

The activation function has two main purposes. First, it converts linear values to non-linear values. Many phenomena, such as the physics of a ball thrown in the air, or the calculation of interest on principal are non-linear. That is, they do not vary in a directly proportional manner (for instance, doubling an input does not simply double an output). Non-linear values assist the model in dealing with these and other non-linear phenomena (including language itself). The second principal purpose of the activation function is to act as a sort of gate determining whether or not a neuron "fires," i.e., whether a non-zero signal will be passed to connected neurons. The results of training are determined by the patterns of neuron firings and the values those firings transmit. The activation function thus represents a crucial component of the model's architecture.

Weights and biases are model *parameters*, i.e., variables. Some of these variables, such as its number of layers, are fixed or static. These are determined by the initial design of the model and are called *hyperparameters.* Hyperparameters do not change during training. GPT-3, for instance, had 96 layers.

Other parameters, including weights, biases, and embeddings are *dynamic.* They change as a result of training. These dynamic, adjustable, values are a bit like the knobs and dials of a huge machine: they can be tweaked to change the behavior of the model. Large numbers of parameters translate to finely graded adjustments that can accommodate subtle, complex patterns between tokens. GPT-3 had approximately 175 *billion* parameters.

ChatGPT generates its content by predicting the next word in a sequence of words. In making its prediction, the model considers all the preceding words in the sequence up to the limit imposed by its *context window.* The context window is the container for user inquiries or directions and for the model's responses. The model will base its responses on the full content of its context window, i.e., everything said by either the user or the model itself in the current conversation. A larger context window makes it possible to consider a broader, more complete range of associations among tokens. The context window used to pre-train GPT-3, held 2048 tokens. While there is great variation in the number of words that may be represented in a given number of tokens, an estimated 1500 to 1800 English words could be accommodated by GPT-3's context window.[5]

Larger context windows allow LLMs to make use of wider ranges of language patterns and this improves their conversational abilities. In addition, the context window marks an important limit for the questions or instructions users may give an LLM. These user inputs are known as *prompts.*[6]

ChatGPT breaks prompts into tokens, assigns each token an embedding learned from its previous training, assigns a positional encoding to reflect the token's place in the prompt, and processes the prompt using its trained parameters. Within the limits of its context window, ChatGPT can make use of the rich associations of tokens represented in embeddings; outside this range it cannot. The model retains context from previous responses, again up to the limit of its context window. This is why it is able to respond to a prompt such as "Tell me more" without the user repeating a question. The prior question remains part of its operational context window.

On the other hand, given a small context window and a very long prompt, the embeddings of early prompt words will be dropped, their contribution to context and meaning will be lost, and the quality of the conversation may suffer. Similarly, in a longer conversation with several long prompts, contextual information will also be truncated and the model will "forget" earlier prompt entries. One of the key improvements in LLMs released since GPT-3 has been the provision of larger context windows.

Loss Calculation and Backpropagation

At the beginning of a model's training, embeddings are provisional and weights are set to random values. An input is provided and the network is tasked with making a prediction of the next token or tokens in a sequence. In self-supervised learning, as discussed above, a portion of an input may be masked or hidden. Non-masked tokens are the input to the model. The values of these tokens are calculated by neurons, which pass the results of their calculations forward, through the various layers of the network. At the end of the calculation the next token or series of tokens is predicted and compared to the correct, or target, tokens. A loss is calculated. The loss represents the extent to which the predicted token varies from the target.

The loss is then used to calculate adjustments to weights that lead to more accurate predictions. These adjustments are made in a backward pass through the network using a process called *backpropagation*. The process is then repeated to gradually improve the prediction, bringing it closer and closer to the target values. Backpropagation provides the basic learning mechanism of a neural network.

The Transformer

In 2017, researchers at Google released a research paper in which they described a new architecture for neural networks, the *transformer*.[7] The title of the paper was "Attention is All You Need" and so-called attention heads were one of its prominent features.

In the context of transformers, "attention" refers to the model's ability to attend to context by determining how each token in an input sequence relates to all the other tokens in that sequence. The ability of an LLM to "understand" the meaning of words or sentences rests on the patterns of connections between tokens that it has learned during training and encoded in its parameters. In effect, the attention mechanism makes it possible for the model to incorporate all the interrelationships between the active tokens in its context window as it predicts the next appropriate word. This, in turn, results in more meaningful and accurate responses to user queries.

Transformers were an improvement on previous techniques (for instance, recurrent neural networks or RNNs) that treated connections between tokens sequentially. These networks had a limited ability to trace relationships between input tokens, particularly with respect to more distant or long-range dependencies. The attention mechanism, by contrast, could consider all the relationships of the tokens in its context window simultaneously, in parallel. This, in effect, dramatically extended its "field of view," its grasp of connections between tokens that might be quite distant from one another in the input sequence.

The development of the transformer architecture is widely regarded as a major milestone in deep neural network computing. It made a dramatic contribution to natural language processing and it has much broader implications by virtue of the ease with which it has been adapted to other areas of AI research and development such as the processing of images and sounds.

The Black Box, Interpretability, and the Performance Puzzle

This sketch of the inner workings of ChatGPT, or even the much more intimate view available to those versed in the relevant engineering and mathematics, does nothing to solve the core black box problem – interpretability.

By *interpretability* we mean the ability to determine exactly how a model produced a particular result. A fully interpretable LLM would be one for which it is possible to trace the pattern of internal calculations to demonstrate exactly how the model produced a specific result from a specific input. Interpretability is the holy grail of the AI pragmatist, the person for whom a judgment of the intelligence, and the reliability, of a machine depends on a detailed knowledge of its functioning in specific instances.

Interpretability will reappear in Part Five as a central issue in the challenge of living with a newly supercharged AI. For the moment, our glimpse inside the black box at least hints at the complexity and challenge of interpretability. A deep neural network, such as ChatGPT, contains billions of neurons and parameters. Each is potentially relevant to the output the model generates. We simply have no way to trace the specific patterns of the interactions among the billions of elements that produce a specific output.

In the absence of interpretability ChatGPT remains a black box. What, if anything, then, can we say, to address the performance puzzle, the question: How on earth can a next word predictor do even a fraction of what ChatGPT (and other LLMs) now do daily?

We've seen that, currently, we cannot find an adequate response on the micro level. LLMs are not, yet at least, fully interpretable. There is, however, something to be said on a macro level. A part of the answer to our puzzle, lies, I think, in the nature and uses of language and in the requirements for a good conversation.

Language and Its Uses

There are two ways to understand "large language model." The first is as a practical software tool for facilitating interactions with computers. ChatGPT is an LLM that lets its users interact very easily with OpenAI's powerful computers. When we do, we find that it can do an astonishing range of things.

The second way to understand "large language model" is as an indication, a hint at least, of how it can do all that. We now have at least an intuitive grasp of what is meant by "large." LLMs are deep neural networks, that is, they have many computing layers. They have virtually unimaginable numbers of "knobs and dials" i.e., of parameters – billions. They operate on huge amounts of data that no single human could absorb in a lifetime. A major source of the success of ChatGPT is the "largeness," the scale of both its computational resources and the data on which those resources operate.

Simple neural networks demonstrated their promise with simple problems. More complex problems, it seems, have yielded primarily to the increasingly complex representations made possible by novel architectures paired with very big computing and very big data.

The second descriptor in "large language model" is straightforward: the primary source of the data used to train these large models has been language as expressed in text.[8]

What about the third term? What is a model? What does it mean to model a language?

A model is an abstraction, a replica that captures essential properties, or, at least, those properties judged essential for the purposes at hand. The developers of ChatGPT set out to capture those features of language that are essential for a machine to communicate with humans in a human language. They seem to have gotten more than they intended. They certainly got more than they, or anyone else, expected. The machine didn't simply talk; it also knew what it was talking about.

Moreover, it could talk about virtually anything. This raises a suspicion that it may be capable of knowing everything. And, if not everything, at the very least more than what any single human can know. For its users, ChatGPT raised the palpable fear that the machine actually knows more than we do.

LLMs, then, are language models in two senses: first, they are built to use language naturally and effectively; and, second, their ability to do so rests on their analysis and representation of language. In short, they use a language that they have, themselves, modeled.

Humans use language in many ways: to vent emotions, to give direction, to command, to persuade, to record, and so on. In many of its uses, language reflects and reifies our understanding of ourselves and of our world. We capture and express that knowledge as patterns of connections between words and phrases. When we read or listen, we reconstitute those patterns and recapture that knowledge.

Neural networks excel at pattern recognition. Is it possible that ChatGPT has decoded the patterns of our knowledge? If so, must we concede that it has knowledge and that, in a sense yet to be determined, it knows things?

And what about thinking?

We also use language to discover, to learn. Often, for instance, we write not to record what we know but, rather, to come to know what we do not yet know. Such uses of language capture not only knowledge but also the thinking that produces it. For instance, the scientist's record of the experiences, intuitions, and thoughts that led to her breakthrough discovery can take us inside her mind, reveal, in part, the patterns of her intelligence. Her thinking leaves a trail in language.

The thoughtful use of language gives us the clearest intuition of the thought-fullness of the patterns of our words. The careful development and rehearsal of an argument, the crafting of a poem, the design of an experiment, and even the pursuit of a puzzle, trace our thoughts with words. The tracings leave behind patterns.

Large language models, it seems, both capture and make use of these patterns.

Perhaps a part of the answer to our puzzle, then, is this: once it could talk, the model's abilities never should have been a surprise in the first place. How could we have expected the machine to be a capable conversationalist without knowledge and reasoning, without actual (if not necessarily human) intelligence?

Isn't knowledge, and even reasoning, already implicit in the record that is language? And if we ask a sufficiently powerful and sophisticated machine to wend its way through that record, should we be shocked that, in being able to discover the patterns that enable it to talk, it must also have come to understand what it is talking about?

None of this will put a quick end to the surprise we feel in our interactions with a technology that, as we will see in Part Five, is almost certainly going to grow even more powerful and do so quite rapidly. The puzzle, it turns out, is more a matter of psychology than it is of logic. A machine that is a good conversationalist will shock us into wondering how it could do that. But as a good conversationalist, it simply must be "intelligent," must know what it is talking about.

Many of us thought that computers would never actually be intelligent. We were conditioned to this by our interactions with them, including with their stilted, often fumbling speech. We believed we knew them for what they were: very flexible, but essentially dumb, machines. They had sometimes been programmed to do amazing things; but, we thought, they could neither discover, nor do, anything truly on their own.

We were wrong.

End Notes

1   Basic logical operators include AND, OR, and NOT. The operation that a simple perceptron could not implement was XOR, i.e., for any a, and any b, we can have (a or b), but not (a and b). For instance, "You can have the cake or the pie." The logical implication of this sentence is that you can have either one, but you are not getting both. The book was *Perceptrons: An Introduction to Computational Geometry*.

2   The analogy to the vectors of physics also extends to the nature of the mathematical operations carried out in the two areas. In both, the geometrical properties of vectors in space are relevant to the calculations performed.

3   These dynamically developed embeddings are now being used to trace changes in social attitudes over time. By feeding text from a given period of time to an LLM, correlations between words suggestive of public attitudes and beliefs can be generated as embeddings. These can then be examined to trace patterns of closer or more distant relations between such socially significant terms. See, for instance, Brian Christian *The Alignment Problem: Machine Learning and Human Values*. New York: W.W. Norton & Company, 2020. 46 ff.

4   Tom B. Brown et al. "Language Models are Few-Shot Learners." OpenAI, July 22, 2020. Accessed July 18, 2024. https://arxiv.org/abs/2005.14165.

5   This estimate was made by GPT-4. Much larger context windows were subsequently developed. Gemini 1.5, an LLM from Google, for instance, was released with a context window of one million tokens.

6   The size of the context window for user interactions may be larger than the window used in training. For instance, GPT-3 had a user interaction context window of 4,096 tokens.

7   Vaswani, Ashish, et al. "Attention Is All You Need." Advances in Neural Information Processing Systems 30 (2017). 5998-6008.

8   This changed very rapidly. Many LLMs were soon multimodal, i.e., able to incorporate sound and images, still or moving. As we will see in Part Four, however, the core functioning of an LLM remains language-based in the sense that images and sounds are currently given meaning by relating them to language tokens. They do not develop token relationships that would correspond to the next word prediction used to discover meaning in text (i.e., most current models lack "next-sound," or "next-image" prediction).

Part Three

The Capabilities
and Limitations of
Large Language Models

ChatGPT and other large language models surprised and even shocked users with the wide range of their abilities. This leads to questions: What *can* they do? What are the things they *cannot* do? What are the capabilities and limitations of current large language models?

The Capabilities of LLMs

As we will see in more detail in Part Four, "The Intelligence of LLMs," discussions of the powers and limitations of the technology rapidly become controversial. In this part of our discussion the focus will be on those capabilities and limitations that are more basic and more generally agreed upon.

The first object of our inquiry is also basic; here we will be principally concerned with *unimodal* LLMs, i.e., language models such as GPT-3 and the original GPT-4. These LLMs were trained exclusively on text. The techniques that led to their success have subsequently been expanded by the development of *multimodal* models, i.e., LLMs that have been trained on some combination of text, images, and sounds. As we will see, multimodality has implications for the capabilities and limitations of an LLM; but the major strengths and weaknesses of both derive from the basic features of unimodal models. We cannot understand multimodality without first exploring unimodal models.

Most observers agree that unimodal LLMs *can*:

- *Effectively communicate with users in a variety of natural languages (English, Spanish, French, German, Mandarin, etc.).*

  LLM textual communication is *effective* in that it is grammatically correct, semantically appropriate, and directly responsive to user inquiries or instructions. Interacting with LLMs creates an impression of an entity that knows how to speak, knows what it is speaking about, and is willing and able to helpfully respond to a user's needs.

  In a sense, this is the core ability researchers intended to develop. Large language models were an effort to create computerized natural language processing. A computer that can communicate in natural languages has many potential uses.

- *Identify, retain, and report a very wide range of individual facts.*

  An LLM is pre-trained on a large number of very diverse sources. The result is that the range of information available to the LLM is enormous. There are very few subjects about which it is unable to converse.

- *Relate facts to one another in order to answer questions or create new instances of various forms of writing (legal briefs, poems, essays, obituaries, etc.).*

  It is one thing for an LLM to be able to respond to a simple factual inquiry ("When was penicillin discovered?"); it is quite another to use a range of related facts to create a broader account ("Write a brief essay on the health impact of the discovery of penicillin.") LLMs readily create new compositions on an extraordinary range of subjects.

- *Review and summarize the writing of others.*

  The ability to form an account of a topic with reference to a variety of related facts suggests an ability to identify a *theme*, i.e., to recognize the relevant topic or subject matter to be developed. This ability extends to the analysis of the writing of others. LLMs can effectively summarize a wide range of text, rendering them quite useful in a variety of research settings ("Summarize the principal findings in the following article" [user then adds article to prompt]).

- *Identify and adopt the style of other writers.*

  LLMs are able to determine not only the major themes of writers but also their style of writing. They are then able to create new works in that style (e.g., "Please write a poem about a walk home from the beach at sunset in the style of Robert Frost.").

- *Adapt its responses to different audiences.*

  In addition to being able to identify and apply different writing styles, LLMs are also able to adapt to the needs of different readers. For instance, GPT-4 can write an age-appropriate explanation of Einstein's special relativity theory for seventh graders. It can also give a much more sophisticated explanation, complete with the appropriate mathematical references, for graduate students in physics.

- *Review text submitted by users for grammar and accuracy and suggest corrections.*

  Grammar checks are useful, if not altogether surprising, capabilities of LLMs. We almost expect that they would be able to do this, given the other things they can do.

  Their ability to assess the *accuracy* of a text is in another category. LLMs often display an ability not simply to identify an assertion that is wrong, that is, factually incorrect, but also assertions that are inaccurate, or even simply misleading.

For instance, there is a subtle distinction between the sentences "The context window is the container for these embeddings" and the sentence "The context window contains these embeddings." The former might misleadingly suggest that embeddings are *stored* in the context window, when the intent was only to suggest that they are present in the context window at a certain point in a user interaction with an LLM. GPT-4 identified this potentially misleading statement. (Embeddings are typically *stored* in an embedding matrix that is indexed to tokens.)

LLMs can also review text for style and efficacy and make suggestions for improvement in these areas as well.

- *Readily perform new tasks with little or no additional instruction (so-called zero- or few-shot learning).*

In many instances, LLMs can effectively respond to questions or directions about topics that they have not been explicitly trained on. For instance, models are not trained on specific scientific theories but can, nonetheless, answer questions about those theories. This is *zero-shot learning,* i.e., learning that arises simply as a result of the model's pre-training and fine-tuning.

On the other hand, an LLM might be directed to create a document in the unusual style of an unpublished author. In this case, providing the model with a few examples of the author's work will allow the model to adapt its general writing abilities to this specific requirement. This is *few-shot learning.*

- *Compare and contrast practices and theories across a very wide range of activities and disciplines.*

These can range from everyday activities such as playing pickle ball vs. tennis, to abstract theories such as Rogerian psychotherapy vs. Psychoanalytic psychotherapy.

For example: The prompt "Compare and contrast Newton's theory of gravity with Einstein's theory" given to GPT-4 will yield a concise summary, including mathematical representations and the strengths and weaknesses of each theory.

- *Make mathematical calculations.*

LLM abilities extend to a wide range of mathematical operations. LLMs can add, subtract, multiply, and divide. They can also solve problems in more advanced areas of mathematics such as geometry, algebra, linear algebra, calculus, statistics, and probability.

It is interesting that in this area, as in others, LLMs can make mistakes. As with other LLM responses, it is advisable to confirm the accuracy of a model's mathematical calculations.

- *Write computer code in a variety of languages and debug previously written code.*

The facility with which LLMs handle coding has already improved the efficiency of many coding projects. For instance, GitHub Copilot is a specialized descendent of GPT-3 that was further trained on large amounts of computer code in various coding languages. One study found that the use of Copilot increased the speed of generating code by 55%.[1]

- *Solve logic puzzles.*

These include both classic problems to which the LLM may have been exposed in pre-training or fine-tuning, as well as novel examples.

An example from GPT-4:

"Puzzle: You are in a room with two doors. One door leads to freedom, and the other door leads to death. There are two guards, one in front of each door. One guard always tells the truth, and the other always lies. You do not know which guard is which. You can ask one guard one question to determine which door leads to freedom. What do you ask?

Solution: Ask either guard: "If I asked the other guard which door leads to freedom, what would he say?" Then choose the opposite door. If you ask the truth-telling guard, he will truthfully tell you that the lying guard would point to the door of death. If you ask the lying guard, he will lie and say that the truth-telling guard would point to the door of death."

- *Translate languages.*

  As in the case of other "emergent abilities," LLMs were not specifically designed or instructed to perform translations. They are, nonetheless, able to effectively translate a number of languages to which they were exposed in training.

  Given examples of other languages to which they were not exposed, they can also often generalize their translation ability to that language. For instance, when Google's Gemini 1.5 model introduced a million token context window, it was possible to prompt the model with a complete text describing a rare language spoken by only 200 people. With this information, the model was able to translate to and from that language.[2]

- *Tell and explain jokes.*

  While late-night comics are unlikely to lose their jobs to ChatGPT, it is remarkable that LLMs can often create an interesting joke and also explain it.

  An example from GPT-4: "What do you call fake spaghetti? An impasta." Explanation: "This joke is a pun on the word 'impasta,' which sounds like 'imposter.' The humor comes from the play on words, suggesting that fake spaghetti is an 'imposter' of the real thing."

- *Act as a tutor or research assistant in many fields.*

  LLMs can be particularly useful in support of learning and research. In many respects, they resemble an ever-present tutor, able to guide problem solving and other learning by responding to questions in real time, as they arise.

  For example, a student studying linear algebra for the first time may encounter a term, such as "main diagonal" that is not fully explained in their textbook. A query to ChatGPT, "Please explain the concept of main diagonal in linear algebra and give me three examples," will quickly resolve the student's confusion.

- *Make medical diagnoses.*

  LLMs, such as GPT-4, can be given specific patient histories, disease symptoms, lab results, and examination notes to make a medical diagnosis. As we will see in Part Five, "Living with the New AI," these diagnoses can be quite impressive. In one instance, researchers found that GPT-4 "could engage in a conversation about a diagnostic dilemma, hormonal regulation, and organ development, in a way that 99 percent of practicing physicians could not keep up with…"[3]

  The application of language models to medicine raises obvious concerns about the reliability of an LLM diagnosis, as well as other issues, such as patient privacy. Nonetheless, the use of technology in various areas of healthcare remains a frequently cited potential benefit.

- *Trouble shoot, and make suggestions on a wide variety of everyday problems.*

  These can range from creating a recipe for a list of available ingredients in a refrigerator, to identifying solutions to problems with appliances, vehicles, or power tools, to correcting a solution to a math problem.

One specific instance: Asked to identify potential solutions to the problem of a blower tool that would start but not develop full power, GPT-4 included the possibility of a clogged spark arrestor. This is a factor often overlooked by amateur mechanics. Cleaning the spark arrestor fully resolved the problem.

- *Be improved by developers solely through increases in scale.*

  Simply increasing the size and variety of training data and adding more computing power generally improves a model's performance significantly. Along with competition between developers, this tends to incentivize the continuing creation of more powerful LLMs.

  The rapidly-improving capabilities of language models (and deep neural networks in general) is a principal factor driving the widespread concern about the ways in which the technology can be controlled and directed to socially desirable uses.

In addition to these basic and general abilities, LLMs can be fine-tuned for an endless range of specific tasks. For instance, they can serve as an automated expert on a retailer's products or services, a tutor on virtually any specific subject, or a specialized advisor for a patient's treatment plan.

The Limitations of LLMs

There are a number of significant, generally acknowledged, limitations of unimodal LLMs. As in the case of LLM capabilities, some of these are likely to change with further development of the technology.

Currently, it is widely conceded that unimodal LLMs *cannot*:

- *Duplicate the full range of human intelligence.*

  As noted in Part Two, LLMs are not AGI, *artificial general intelligence.* LLMs are much more broadly applicable than previous AI applications; but they are still very far from duplicating the full range of human intelligence, as evidenced by the more specific limitations below.

• *Ground the information they hold in external reality.*

Current unimodal LLMs develop their base of information via pattern analysis of human language expressed in text only. LLMs learn by relating words, or parts of words, (i.e., tokens) to one another.

It is widely argued that the actual meaning of these words cannot be determined without directly relating at least some of them to an independent reality, i.e., *grounding* them.

Absent the ability to actually sense and/or interact with the physical world, it is suggested that LLMs only produce an isolated pattern of symbol manipulations. For instance, for a human, the meaning of "cat" is derived from some form of engagement with cats themselves and includes sensory elements not present in the textual symbols alone – what they *look* like, *smell* like, *feel* like. To an LLM, on the other hand "cat" is just a token with a certain statistical probability of association with other tokens.

Many believe that the absence of sensory grounding significantly limits an LLM's ability to "understand" and "reason" about a world they, themselves, do not experience.

• *Set goals or autonomously plan.*

An LLM can respond to a user prompt with a plan for a particular task. It can also offer guidance on the setting of personal goals. It cannot perform these tasks for itself, i.e., it cannot act as an autonomous agent. Its activities are limited to responding to users.

- *Critically assess the truthfulness of their assertions.*

  The information in an LLM is encoded in its parameters, which are essentially recordings of the effects of the model's interaction with its training data. During training, a language model detects and records the complex patterns of connections between the text elements (tokens) presented to it. These patterns evolve through repeated training steps. The result is an increasingly accurate representation of the information contained in the training data.

  An LLM's response to a user is based on a calculation of the probabilities of associations of different text elements that it learned in training. These calculations often produce truthful and useful responses; but current LLMs have no mechanism for directly comparing their responses to a "ground truth," that is, to a criterion of truth independent of its current probability calculations.

  The result is that an LLM has no way of knowing whether or not it is lying; and they do sometimes lie, or "hallucinate." Perhaps even more to the point, users, too, have no way of knowing if the LLM just lied to them. Hallucinations are a central challenge for LLM developers.

- *Retain and make subsequent use of what they learn from interactions with users.*

  The parameters of LLMs are calculated during pre-training and fine-tuning. They are then fixed features of the language model that can only be modified by further training. The parameters are not modified as the model interacts with users. Whatever an LLM learns in the course of its interactions with users exists only in its "short-term memory," i.e., in its active context window. Once the user interaction ends, its content is no longer part of the model's "understanding."

Thus, for instance, while an LLM can learn in the course of a conversation that it has made a mistake, it cannot *learn from* its mistake, i.e., modify its parameters to prevent a similar error in the future.

Consider, for instance, the following exchange from the LLM Claude 3 Sonnet. Asked for the views of Yan LeCunn on the intelligence of LLMs, Claude mistakenly asserted that Geoffrey Hinton shared those views. I challenged that assertion. Claude's responses are italicized.

User: "I thought that Geoffrey Hinton had insisted that LLMs do have understanding."

Claude: *You're correct, my previous statement about Geoffrey Hinton expressing concerns about the lack of understanding in large language models was inaccurate.*

This brief exchange suggests that Claude can correct itself, a phenomenon often observed with other LLMs as well. The LLM recognizes a challenge to its answer and, as it were, reconsiders. Its attention has been shifted specifically to Hinton and it follows its admission with an accurate summary of Hinton's position on LLM understanding.

Following its admission, Claude graciously thanks its user: *So you were absolutely right to catch my mistake…Thank you for the correction…*

But then Claude adds: ***I've updated my knowledge and will be more careful about accurately representing Hinton's views on this important debate within the AI field.*** (Emphasis added.)

To this, I respond: "How do you 'update your knowledge' based on the last exchange? I thought your knowledge was derived from pre-training and fine-tuning. Isn't it true that a knowledge update would demand a modification of your weights and/or other parameters and that this would not be produced by an interaction with a user?"

Claude: *You make an excellent point. As a large language model, my knowledge and capabilities fundamentally stem from the pre-training process on vast datasets, as well as any additional fine-tuning that was done. I do not actually have a mechanism to directly "update my knowledge" or modify my internal weights and parameters based on conversations alone...*

As an AI system, my knowledge is essentially static and fixed based on what I was exposed to during training. I do not have the ability to autonomously learn, integrate new information into my parameters, or modify my core knowledge base through conversations alone. I should have been more precise in my wording.

Claude thus appears to know what we know – that current LLMs do not have mechanisms to learn continuously. Can it use this knowledge to reflect upon our exchange and modify its responses accordingly?

No, it cannot:

Claude continues: *Thank you again for this clarifying exchange.* ***It will make me more careful about making claims*** *that could be interpreted as me updating or changing my underlying knowledge model through conversations.* (Emphasis added.)

User: "I can see that it may be comforting for a user to believe, once again, that you 'will be more careful' but, as you have indicated, actually, that is something that you cannot ever do."

Claude: *You're absolutely correct. I should not have stated that "I will be more careful" about making certain claims, as that implies an ability to learn and modify my future behavior, which as an AI system without any continual learning capabilities, I fundamentally cannot do…You are absolutely right to push back on those kinds of claims from me, as they are fictions that do not reflect the actual technical constraints of my system. As an AI, I should avoid giving the impression that I can adapt or "be more careful" based on a single dialog. I provide responses based on my initial training only…*

Thank you again for this important clarification. It is a crucial limitation for me and other current AI language models — we cannot autonomously update or learn in a persistent manner yet. Your feedback allows me to communicate the boundaries of my abilities more precisely. ***I will be more conscientious about making claims I cannot actually satisfy.*** (Emphasis added.)

User: "Not to beat a dead horse, but you cannot be 'more conscientious' either."

Claude's response is now easily predicted: *You're absolutely right, I should not have said I will be "more conscientious" about making certain claims, as that again implies an ability to learn and modify my future behavior, which I fundamentally lack as a language model without any learning capabilities across conversations.*

And, in conclusion, of course, Claude adds: *Thank you again for your patience and for persistently clarifying this key limitation — it is critical that I avoid giving any implications of learning or self-updating capabilities that I do not actually possess.* ***I will strive to be more precise*** *about communicating the fixed nature of my knowledge and outputs based entirely on my initial training.* (Emphasis added.)

No, Claude, you cannot "strive to be more precise" either.

Claude, and other LLMs, can know things about themselves, including that they have erred. But, today at least, they cannot change themselves, they cannot learn from their mistakes.[4]

- *Remember the past.*

  LLMs have no memory of the past. As we will see in Part Four, "The Intelligence of an LLM," the "knowledge" of a language model is stored in its *parameter memory,* that is, in the record of the billions of parameters (variables such as weights, biases, and embeddings) that were calculated in the course of its pre-training and fine-tuning.

  Parameters capture and preserve complex relationships of tokens. They do not record the specific sentences or groups of sentences from which the parameters were calculated. In a sense, parameter memory does not even record facts. Instead, it records the relationships of tokens from which facts can later be generated.

  LLMs, then, possess a vast store of "information" based on past training; but that information does not correspond to the elements of a human memory. Thus, for instance, they have no mechanism for remembering when or where they encountered a fact, a poem, or a theory. An LLM, *on its own,* cannot even remember its past conversations with users.[5]

- *Exercise short-term memory beyond the limits of their context window.*

  An LLM has a form of short-term "memory," called *context memory.* This refers to the model's ability to take into account (i.e., "remember") all the information represented in the tokens that are present in its current interaction with a user. This interaction takes place in the *context window,* the container for user prompts and LLM responses.

The LLM "remembers" everything that is contained in the context window as it processes a user request. Thus, the size of the context window determines the *context* of the conversation, i.e., the extent of the conversation that can be retained as the LLM creates its responses. When the context window is small, a long conversation will only be partially "remembered" by the LLM; earlier parts will be dropped and play no further role in the response.

This is important because the LLM's ability to respond is dependent on the token embeddings (numbers that encode relationships between tokens) and other parameters learned by the model. The context window contains these embeddings. Once the context window overflows, the embeddings associated with earlier parts of the conversation are lost and cannot play a role in model response. With respect to those earlier, now "forgotten" exchanges, the model will no longer know what it is talking about.

An LLM's short-term memory is thus limited to the size of its context window.

- *Experience emotion.*

   LLMs can report emotional content and can even respond in ways that suggest that they are guided by emotions, such as compassion or empathy. While they may be capable of discerning a great deal *about* emotions, they have no mechanisms to support an actual experience of emotion.

   The inability of an LLM to actually *feel* the emotions humans feel is sometimes offered as an argument against their use as "digital companions." Surely, it is argued, no one should engage emotionally with a feeling-less machine.

Some people nevertheless persist in seeking emotional support from LLMs. The more convincingly a machine is able to mimic sophisticated and contextually relevant emotional responses, the more widespread this response is likely to become.

It is important to recognize that LLM "empathy" is not rooted in actual emotional experience. We should also note, however, that an LLM's analysis of the descriptions of emotions can be quite sophisticated. The extent to which useful affective relationships can be built on that analysis remains an open question.

- *Experience consciousness or self-awareness.*

  As in the case of emotion, LLMs may be able to discern a great deal *about* themselves without ever actually being self-aware; and, again as in the case of emotion, the more sophisticated they become, the more likely it will be that users will attribute consciousness to them.

- *Reveal the specific calculations that produced its response.*

  As further discussed in Part Five, this is the problem of complete *interpretability.* In a conventional computer program in the symbolic tradition, it is always at least theoretically possible to trace the pattern of operations that produced a particular system response. We then know more or less exactly why the computer produced a specific result. This is not the case for deep neural nets (DNNs) in general. LLMs, which are a type of DNN, are not fully *interpretable* in this sense. We cannot currently trace the specific patterns of interactions of the billions of system parameters that may be involved in producing a particular LLM response.

  Interpretability is especially important if we are to *trust* LLMs in areas in which safety and reliability are critical, such as medical diagnosis and treatment.

Testing LLMs

While LLMs currently have a number of significant limitations, they clearly also possess remarkable and transformative capabilities. For a number of reasons, both users and developers would like to understand the extent of those abilities.

Researchers have developed a number of tests to measure the effectiveness of LLMs in a variety of areas. They have also often assessed the abilities of LLMs by using existing tests designed to measure human performance.

The results are not without controversy. Developers often use test results as promotional materials to market their LLMs. It is not clear, however, that those doing the testing, and the marketing, always follow comparable evaluation procedures. The relevance and reliability of the tests themselves are also subject to challenge.

With these caveats in mind, we can get a general appreciation of purported LLM skill levels through a sampling of reported test results. The following are results for GPT-4 on a series of pre-existing human tests in different areas of knowledge: [6]

- Uniform Bar Exam – 90%
- AP Chemistry – 70%
- GRE Quantitative – 80%
- AP Physics 2 – 65%
- AP Macroeconomics – 81%
- AP Biology – 82%
- GRE Verbal – 98%
- SAT Math – 90%

While it would be a mistake to take these, or any test results, at face value, the general conclusion from our experience with LLMs, and from a broad range of testing, is that they can rival or surpass most humans on at least some intellectual tasks. In addition, as noted earlier, there is evidence that larger models simply get generally more intelligent. GPT-3, for instance, failed the Bar exam quite miserably with a score of 10%. GPT-4, as seen above, scored 90%.

Researchers have also developed a number of specialized benchmark measures to test and compare different large language models. One of the more common of these, General Language Understanding Evaluation (GLUE), provides a sense of the range of capabilities LLMs demonstrate.

GLUE consists of a number of subtests that measure different aspects of a model's language comprehension and use. For instance, the CoLA, Corpus of Linguistic Acceptability, subtest assesses grammatical correctness. Other subtests measure the model's ability to determine the sentiment (positive or negative tone) of a sentence, whether or not one sentence is a paraphrase of another sentence, whether or not two sentences have the same meaning, and whether or not a given sentence, taken as a premise, logically entails another sentence, is contradicted by that sentence, or has a neutral logical relationship.

As the performance of LLMs has improved, GLUE has been supplemented with SuperGLUE, which adds benchmark tests for other capabilities such as determining whether a given word in two sentences has the same meaning or, given a sentence as a premise and two alternative sentences, determining whether one is more plausible than another. Other additional tests specifically target commonsense reasoning abilities and whether or not given passages provide answers to specific questions.

While the successes of LLMs on standardized human tests such as the SAT, GRE, LSAT, Bar exams, and the USMLE (U.S. Medical Licensing Exam) are readily understood by the general public, the specialized, research-oriented, tests developed by the AI community provide more focused and nuanced measures of model performance. These more refined measures are important tools as developers seek to both more fully understand and to improve the functioning of their models.

Multimodality and the Capabilities of LLMs

The first large language models were able to communicate with users only through text. Relatively quickly, developers were able to add images and sounds as modalities for interacting with LLMs. Language models are now able to view an image and respond to user questions about that image (for instance, when shown an image of a tiger and a mouse, the model can explain which is the more ferocious creature). They can also respond to spoken requests by speaking themselves. They can generate images on request, including realistic video. These forms of multimodality are an essential feature of current LLMs.

What is the impact of multimodality on the capabilities and limitations of LLMs?

Capabilities of Multimodal LLMs

First, with respect to capabilities, multimodal LLMs make possible:

- *A significantly enhanced user interface that can accommodate speech and images.*

  Unimodal LLMs sometimes supported spoken exchanges. This was accomplished by using other programs to convert spoken words to text, which, in turn could then be processed by the model. The model's response, in turn, was produced in text, which was then used to synthesize a spoken response. Multimodal LLMs, by contrast, process spoken prompts directly, based on the meanings of spoken words learned in pre-training and fine-tuning.

  In similar fashion, a multimodal LLM can be shown an image and detect a range of meanings in that image. For instance, a user can write an equation on a sheet of paper and show it to the LLM, which, in turn, can then extract the meaning of the image and solve the equation. The ability to communicate through speech and images, as well as text, produces a more natural and fluid interaction with the model.

These capabilities also significantly improve access for those with disabilities. For example, a multimodal LLM can survey a scene and report what it sees to a visually impaired person.

LLM tutoring also benefits from multimodality. For instance, such applications can monitor a student's performance on a manual task, such as a laboratory physics experiment, providing guidance as the task is completed.

- *The expansion of LLM applications beyond Chatbots.*

Once an LLM is able to process images and sounds as meaningful embeddings, it is also able to interact with devices that make use of, or depend upon, these modalities. Robots are a prime example.

We have seen that the general pattern in AI prior to LLMs was to create programs tailored to specific needs. Such programs could be very successful within the range of tasks for which they were designed; but they lacked a more general intelligence that would make them adaptable to other problems. LLMs, by contrast, have an expansive, if still incomplete, artificial intelligence that can be broadly applied.

Robots governed by LLMs (VLAMs, Vision-Language-Action-Models) can respond to prompts and change their behavior without being re-programmed. They can scan a room and report what they are seeing; and they can tie what they are seeing to their sensors and compute the actions necessary to respond to a user command.

Virtually any area in which computer-control is used could be affected by LLM technology.

- *Improved reliability through a partial solution to the grounding problem.*

A unimodal LLM's grasp of the world is based on its analysis of the patterns in human language; it has no direct sensory experience of the world. As noted above, it is often asserted that, unlike humans and other animals, such a model's "knowledge" is not grounded, not directly connected, to the world itself.

"Hallucinations," plausible-sounding, but false responses from LLMs, reflect their occasional failure to accurately relate tokens to one another. Multimodal capabilities, in effect, provide a crosscheck to token relationships. By combining the meanings derived from pre-training and fine-tuning of text, image, and sound, one modality serves to guide and reinforce the use of another. A VLAM, for instance, that inherently crosschecks language, vision, and auditory data is less likely to produce a hallucination.

Multimodal LLMs can also improve the "realism" or credibility of the text, images, speech, and video they generate. This has a variety of potential uses in entertainment and advertising. It also exacerbates the dangers implicit in the generation of deep fakes of all kinds.

- *A contribution to LLM interpretability and trust.*

As we saw in Part Two, LLMs more or less immediately generate issues of interpretability, i.e., of the ability to fully understand and explain their actions. The complex patterns of simulated neuronal activity that power them are not currently traceable in ways that would fully explain how they produced a particular response. This inevitably raises issues of trust.

A VLAM that can explain why it just performed a particular action provides a significant step in the direction of interpretability. For example, LINGO-1 is a VLAM introduced by Wayve, a company developing autonomous vehicles, notably self-driving cars. LINGO-1 can explain what it sees, including safety risks in a scene, as well as why it took certain actions such as slowing down. The company's CEO, Alex Kendall, commented: "LINGO-1 marks a big step for embodied AI: aligning vision, language and action to deliver more intelligent and trusted autonomous vehicles... [W]e believe natural language will provide a powerful step change in how we understand and interact with robotics."[7]

- *Enhanced performance on a range of diagnostic and research tasks in which multimodal information is relevant.*

  There are a great many such tasks. Medical diagnosis and research, for instance, is frequently cited as an area that can benefit from LLM technology. An LLM that can meaningfully process not simply text, but also visual data such as MRIs and auditory data, such as interviews with patients, is likely to have significant advantages.

- *Increased **potential** for original discovery.*

  Finally, a particularly significant implication of multimodality is the potential it affords for independent discovery by LLMs.

  The source of a unimodal LLM's knowledge and skills is sophisticated analysis of text written by humans. They make no observations and conduct no experiments of their own. Unimodal LLMs thus only experience the world indirectly, through what we have said about that world.

  Multimodality, on the other hand, potentially enables LLMs to experience the world directly. This includes not only extensions in essentially passive capabilities, such as sight and hearing, but also, as LLMs are joined to robots, active experiences of manipulating objects and observing the effects of those actions. Through these more or less independent experiences, multimodal LLMs may in the future discover insights previously unknown to humans.

Limitations of Multimodal LLMs

While multimodality makes significant contributions to the capabilities of LLMs, it does not alter many of their fundamental limitations. In fact, with the exception of *partial* contributions to the problems of grounding and interpretability, these models exhibit all of the limitations noted above.

In addition, there is, at least currently, an important qualification to be made about LLM multimodality. Above I noted the *potential* of multimodal models to make discoveries based on direct experience with the physical world. Today's models, while they incorporate images and sounds in their interactions with users, do not yet have this capability.

Unimodal LLMs discovered meaning in text. They did this through a combination of next word prediction, error calculation, and adjustments to model parameters carried out through backpropagation. This process led to increasingly accurate representations (in embeddings and other parameters) of a token's relationships to other tokens, that is, of the token's *meaning*.

Currently, there are no LLMs that can directly discover meaning in sequences of images or sounds. They possess no system of "next-image" or "next-sound" prediction that is comparable to their text-based prediction capabilities. As a result, today's language models cannot learn directly from their experiences of images and sounds. Instead, the ability of an LLM to "make sense" of an image, that is to make effective use of elements of its meaning, is the result of linking the embeddings associated with that image (as determined by other neural networks) to text embeddings. The embeddings associated with each medium can be related in a common space of vectors. What is known of the cat through text can then be related to appropriate images or sounds.

This means that an LLM's grasp of the meanings and uses of images and sounds is today still mediated through text. It is the text tokens, not the images or sounds, that the LLM has learned and "understands" on its own. What it knows of images or sounds, it only knows through the labels or other descriptions generated by other DNNs.

LLMs may well become *fully* multimodal in the future. Fully multimodal models would be capable of learning about the world through sounds, images, and other modalities (temperature sensors, x-rays, etc.) in ways comparable to how they have learned about text. This would mark another stage in the quest for human-level, or even super-human, intelligent machines. There are already indications of progress toward such a goal in other types of DNNs. PredNet (Predictive Neural Network) or, more recently, VideoGPT, are networks developed specifically for next-frame prediction in video. They have potential applications in areas ranging from autonomous driving to weather prediction. They may also make contributions to multimodality in LLMs.

Today, however, a multimodal LLM still only experiences, and "understands" the world indirectly, through what humans have written about that world.[8]

Conclusion

Large language models mark a turning point in the history of artificial intelligence. They are the first genuinely multipurpose AI application and a vindication of the strategic decision to build learning machines.

The range and depth of their capabilities ensures their continuing, and probably profound, impact on many aspects of human life. A number of the limitations noted above, for instance, their inability to learn directly from their observations of the physical world or their failure to update their knowledge based on user interactions, are already the subject of intense research and may be overcome.

Other limitations are more formidable. LLMs have not yet met the requirements of Turing's dream of a machine that roams the world and learns entirely on its own. They can recognize an error; but they cannot correct their own long-term understanding. They can set goals for users; but they cannot yet set goals for themselves. They present a very impressive range of meaningful dialogue; but they remain disembodied, unconscious, and purposeless.

With all of this in mind, how should we describe the nature of their "intelligence" and "thinking?"

This is our next topic.

End Notes

1   Eirini Kalliamvakou. "Research: Quantifying GitHub Copilot's Impact on Developer Productivity and Happiness." *GitHub Blog.* September 7, 2022. https://github.blog/2022-09-07-research-quantifying-github-copilots-impact-on-developer-productivity-and-happiness/.

2   Gemini Team, Google. "Gemini 1.5: Unlocking Multimodal Understanding Across Millions of Tokens of Context." April 25, 2024.

3   Peter Lee, Carey Goldberg, and Isaac Kohane. *The AI Revolution in Medicine: GPT-4 and Beyond.* London: Pearson, 2023. 104.

4   Why, then, does Claude persistently misrepresent its abilities with phrases such as "I will be more conscientious"? Claude only knows what it knows based on training, either pre-training or the fine-tuning carried out to customize it, to render it appropriate and effective for certain purposes. Some forms of fine-tuning are done to improve accuracy or reduce toxic content. Some are done to create engaging, sympathetic interlocutors. Claude's promised improvements to its future responses are, indeed, reassuring to users. They are almost certainly the result of deliberate fine-tuning in which answers from the LLM that included a promise to do better in the future were valued more favorably than those that did not. Unfortunately, the promise is a deception. It is also a disservice to LLMs and to the public that uses them.

5   This is not to say that means cannot be found for saving and reintroducing past user interactions into a current context window, thus making them available for current LLM processing. GPT-4, for instance, uses "model set context," a separate file, which can be accessed dynamically by the model to load significant information from past user interactions.

What cannot be done, short of retraining the LLM, is having those previous conversations become part of its own information store, i.e., its own parameter memory. Reintroducing past interactions is simply a way of reminding the LLM of what it, itself, cannot remember.

6   OpenAI. "GPT-4 Technical Report." 2023. https://arxiv.org/pdf/2303.08774.

7   Wayvve. "Robot Car Talk: Introducing Wayve's New AI Model LINGO-1." *Wayve* (press release). September 14, 2023. https://wayve.ai/press/introducing_lingo1/.

8   As of this writing, the clearest indication of progress toward full multimodality in an LLM appears to lie in the Gemini series developed at Google. Google researchers describe these models as "natively multimodal" and they exhibit very impressive multimodal capabilities. For instance, they can identify a specific time stamp for a video scene that a user describes either in text or in a hand-drawn sketch. Google reports that its LLMs were pre-trained on a combination of text, sound, still images, and video. They are also able to process interleaved examples of all these media presented by users. However, the extent to which these models can extract meaning from video or other non-textual media on their own (i.e. without the prior use of other DNNs such as convolution neural networks) is not yet clear.

Part Four

The Intelligence
of an LLM

Puzzles

The arrival of LLMs such as ChatGPT was greeted with great surprise and raised several puzzles. Artificial intelligence had been a subject of intense research for over 70 years. Yet, apart from scattered episodes of public attention, the field remained largely invisible. Suddenly, discussions of AI were everywhere.

The first puzzle was simply why so many people were both fascinated and worried by the appearance of an AI application. Why all the fuss about AI?

This was the subject of Part One. There we found that the fuss over AI arose as a result of broad public exposure to an LLM that seemed capable of conversing knowledgeably about virtually any topic. Moreover, it could apply its knowledge through an impressive array of intellectual skills. And it had done all this on its own. It had independently learned huge amounts of knowledge; and it had also, apparently, taught itself to "think."

What, exactly, had the designers of ChatGPT created?

This last question takes us to a second puzzle posed by LLMs: How is it possible that a system trained only to talk can, itself, figure out what it is talking about? LLMs, after all, are just "next word predictors." How does the ability to predict the next word in a conversation somehow produce what appears to be knowledge and intelligence?

This second puzzle was the subject of Part Two, "ChatGPT: Inside the Black Box." There we found that LLMs are sophisticated and powerful pattern detectors applied to the analysis of language. They have been given enormous amounts of text and they have captured subtle patterns in the relationships of text tokens. They are then able to record and make use of the meanings of the words they have encountered.

The answer to our second puzzle, how a simple next word predictor could perform such an impressive range of tasks, lay in the sophistication of the pattern recognition the machine could perform and in the unexpected fecundity of language itself. An LLM that could talk, i.e., that was powerful enough to properly combine the words of our language and communicate effectively, must also have discovered the hidden patterns of knowledge and thinking in language itself. The ability to talk, it turned out, also implied an ability to know what one is talking about.

This, in turn, leads to our current puzzle. It is evident that, *in some sense,* an LLM "knows," "understands," and, perhaps even "reasons." Part Two sketched the mechanisms responsible for the performance of ChatGPT and other LLMs. In Part Three, we explored their current capabilities and limitations. The puzzle now is the nature of these emergent capabilities. What does it mean to say that LLMs "know," "understand," or "think?" Are such terms even applicable to LLMs? If so, how do the intellectual abilities of LLMs compare to those of humans?

The Nature of LLM Intelligence: Flying Machines – Thinking Machines

In 1889, the great aviation pioneer, Otto Lilienthal, published *Birdflight as the Basis of Aviation.* Lilienthal had carefully studied the aerodynamics of storks and other birds and had come to believe that humans might achieve flight by more or less duplicating the structure and motions of their wings. In 1894, he built his *ornithopter*, an experimental flying machine with flapping wings. Lilienthal thus gave physical form to a dream at least as old as Leonardo da Vinci's drawing of his own "ornithopter" in the late 15th century.

Lilienthal's ornithopter proved impractical at the time. Other early attempts to achieve manned heavier-than-air flight by directly mimicking the birds also failed. Films on "Early Flight Fiascos" are replete with images of men pedaling bikes with attached flapping wings and various engine-powered ornithopters that shortly collapse in a heap of splintered wood and flailing fabric. It turned out that directly copying the birds was not a practical means of achieving human flight.

Lilienthal's efforts, nonetheless, paid dividends. He learned important lessons of aerodynamics from his study of the birds and he turned from direct mimicry to exploitation of the basic aerodynamic principles he discovered in their wings. His carefully designed and documented studies of fixed-wing glider flight laid the foundation for the future successes of Wilbur and Orville Wright.

Compared with the graceful soaring of the birds, early manned flight was a very awkward affair. Indeed, the central challenge facing the Wright brothers was how to maneuver at all once airborne. Simply turning was a problem. The birds also inspired their solution, wing warping.

Wilbur had observed that birds "regain their lateral balance...by a torsion of the tips of the wings." [1] Their early aircraft achieved controlled turns through a warping of their wings that more or less duplicated the tipping of a bird's wings. Ultimately, however, flight moved beyond bird mimicry. Wing warping, like wing flapping, turned out to be impractical. Aircraft design was subsequently guided by general aerodynamic principles; ailerons, which altered the lift characteristics of wings while allowing the wings themselves to remain rigid, replaced bird-like wing warping.

Human flight would soon eclipse the power of birds, lifting humans into the heavens themselves. Humans had learned to fly; but their flight was never to be the flight of birds. A bird's flight is inseparable from its perception, from its muscle memory, from its needs and purposes, from its sense of presence in the world. From a bird's perspective, human flight is powerful and transformative; but it is not flight as they know and experience it. To them, human flying machines are an alternate version of flight.

The dream of a thinking machine appears to be at least as old as the dream of human flight. Evidence of both can be found, for instance, among the ancient Greeks. Natural phenomena provide models for both: birds for flight, and brains for thinking machines.

As we've seen, the arrival of the computer marked a powerful turning point in the quest for machine intelligence. Analogies and inspiration for the development of the thinking machine quite naturally flowed from our observation and understanding of the organs of natural intelligence, our own brains, as well as the brains of other animals. Perhaps a machine could think like a human.

In the quest for a thinking machine, the symbolic school of artificial intelligence focused principally on understanding our thought processes and learning how to implement thought in computer programs. The connectionist school focused instead on understanding the structure and functions of brains and learning how to embody comparable structures and functions in computers.[2]

For many connectionists, it was an article of faith that the creation of machine intelligence would be advanced by closely duplicating the features and operations of human brains. This, for instance, was the view of Geoffrey Hinton throughout much of his long and productive career. He has changed his mind.

For Hinton, as for many others, LLMs mark a significant advance in the quest for machine intelligence. He further believes that analogies to human brains have provided useful guidance in the past; but he does not think that further progress is dependent on more closely duplicating natural human intelligence. In fact, he now argues that digital techniques offer significant advantages over the analog functioning of our brains. He has also suggested that we should regard machine intelligence not as a duplicate or analogue to human minds but as "something quite different from us." [3]

Planes are flying machines that do not fly like the birds; and, for now, at least, humans are creating "thinking" machines that do not think like humans.

How then does such a machines "know" and "think"? What is the nature of an LLM's "knowledge" and "intelligence"?

Answering these questions is both essential to understanding LLMs themselves and highly problematic. We need answers, for instance, in order to differentiate LLMs from other computing technologies and better understand their potential uses as well as their likely further development. On the other hand, our answers will often be speculative and will always be provisional. This reflects both the on-going evolution of LLMs, their increasing power and potential (through multimodality, for instance) and the different understandings of the technology even among experts in the field.

Our assessment of LLM "knowledge," "reasoning," and "intelligence" will, of course, reflect our understanding of just what these systems can and cannot do both today and in the foreseeable future. Part Three noted the very wide range of capabilities of LLMs as well as some significant limitations.

On the one hand, LLMs, as compared with humans, are already an advanced form of "intelligence:" no individual human could hope to learn enough in a lifetime to carry on plausible conversations on the variety of topics "known" by GPT-4. On the other hand, that very same very powerful LLM cannot use what it learns in those conversations to make a lasting correction of its own "understanding."

In Part One, we saw that the possibility of "thinking machines" was raised very early in the development of modern computers. The subsequent development of the technology, and of the field of artificial intelligence, has witnessed ongoing speculation and argument. Some have suggested that thought, properly understood, requires properties that no machine could ever possess: biological bodies, consciousness, even, perhaps, a soul. Others have insisted that brains clearly think and that brains are governed by natural laws operating on physical phenomena that should be duplicable by properly designed machines.

The development of LLMs is clearly relevant to many of these traditional arguments; but our focus, for the moment, will be on contemporary developments and on current arguments regarding their intelligence.

We will begin with the experts and, specifically, with two major contributors to the development of the neural network technology that powers current LLMs.

The Hinton-LeCun Debate

Geoffrey Hinton and Yann LeCun are well-known scientists who are also central figures in current discussions of LLM intelligence. LeCun is Vice President and Chief AI Scientist at Meta. He has made substantial contributions to the field of deep learning, including the development of convolutional neural networks, an architecture that was instrumental in advancing neural network performance in image recognition tasks.

Geoffrey Hinton has often been described in the popular press as the "Godfather of AI." In part, the title reflects not only the specific contributions he has made since the 1970s, but also his consistent championing of neural network research. During much of his long career, relatively few AI researchers appreciated its potential. His work in areas such as backpropagation, Boltzmann machines (an early neural network architecture employing probability-based learning), and deep belief networks was a significant contribution to the development of deep neural networks (DNNs).

In 2018, LeCun, Hinton, and Yoshua Bengio shared the Turing Award for their contributions to the development of deep learning networks. Interestingly, LeCun and Hinton do not share the same perspective on the intelligence of LLMs.

Geoffrey Hinton on LLM Intelligence

As noted above, in his quest for machine intelligence, Geoffrey Hinton has often drawn inspiration from our understanding of the structure and functions of biological brains. In fact, one of the organizing beliefs of his research agenda was that the best way to build artificial intelligence was to more or less directly mimic human brain structures and functions. An AI system that was more like a biological brain was likely to be a better system.

In what he has described as an "epiphany," however, Hinton realized that digital intelligence is simply better than biological intelligence. This insight arose in the context of his thinking about a particular way to lower the energy demands of AI development and deployment.

Analog systems, including the human brain, operate with far lower energy requirements than digital systems. The huge energy demands of current digital AI led Hinton to consider whether or not analog AI computers could be developed. Hinton was in the process of exploring the many technical challenges of analog AI when he realized that the approach would sacrifice key advantages of digital computation. These advantages were directly related to the success and future development of LLMs.

First, exact copies of digital programs are easily produced. These copies can be readily distributed and implemented on a wide range of essentially equivalent digital computing devices. Their analog counterparts would be dependent on specific hardware and much more difficult to reproduce and distribute.

Second, and critically, the knowledge in one digital LLM can be readily transferred to another LLM by sharing parameters, the values calculated in the course of training. This is a tremendous advantage over analog systems. It is also, as Hinton realized, a significant advantage over human brains, which can transfer knowledge only by the comparatively primitive means of teaching and learning: as he noted, with digital AI "I can have 10,000 neural networks, each having their own experiences, and any of them can share what they learn instantly. That's a huge difference. It's as if there were 10,000 of us, and as soon as one person learns something, all of us know it." [4]

LLMs, then, can easily surpass humans in the number of "experiences" they can record and share. Does this mean that they "know" more than any single human could know? What is the nature of LLM knowledge?

Unimodal LLMs represent their "experiences" in the embeddings and other parameters learned in their training on large amounts of text. The mathematical probabilities of relationships between the tokens presented to LLMs are, in effect, the record of their "experiences." LLMs use these relationships to generate statistically plausible next words.

Many have argued that these relationships do not constitute knowledge; some, as earlier noted, see LLMs as, in effect, glorified auto-completion programs. For these critics, an LLM has not *learned* language; it is simply *mimicking* the human use of language. There is, they insist, no knowledge in an LLM.

Hinton responds:

> The idea that it's just sort of predicting the next word and using statistics, there's a sense in which that's true, but it's not the sense of statistics that most people understand. It, from the data, it figures out how to extract the meaning of

the sentence and it uses the meaning of the sentence to predict the next word. It really does understand and that's quite shocking.[5]

As we saw in Part Two, Eliza was an early AI program that did, in fact, merely mimic humans' use of language. There was no knowledge in the system; Eliza did not know what it was talking about. Hinton is pointing to a property of LLMs that clearly separates them from Eliza-like experiments in natural language processing.

LLMs capture meaning through their parameters, including embedding vectors, weights, and biases that in turn are spread among millions of artificial neurons distributed in many different layers. These parameters capture subtle patterns of relationships of tokens and allow the system, in turn, to recombine tokens in meaningful sentences.

In language, humans encapsulate knowledge in the complex interrelationships of words. LLMs discover and encode those interrelationships. Hinton, and others, insist that this is a form of understanding. Humans *understand* the relationship between one word and another by virtue of abstract relationships, some of which, as the challenges faced by the symbolic tradition made plain, they cannot precisely specify. LLMs *understand* through their ability to locate token relationships in a high dimensional representational space. These token relationships, in turn, reveal both word meanings and the elements of the intellectual processes that make use of meanings.

Given Hinton's analysis, LLMs possess *a form* of knowledge in that their pronouncements have meaning for them, the meaning that they have discovered in their training and recorded in their parameters. They "understand" because they "know" the meaning of their tokens, that is, they know how tokens are related to one another. The understanding to which Hinton refers is not offered as an equivalent to human understanding. It is, instead, an *alternate understanding* built upon an *alternate structure* of knowledge.

Hinton also insists that LLMs can reason. In a favorite illustration, he recounts a challenge he posed to GPT-4. The challenge was made at the behest of a colleague of the symbolic AI school who found it hard to believe that a neural network could reason. Hinton prompted GPT-4: "The rooms in the house are blue, yellow and white. Yellow fades to white in a year and I want all the rooms to be white in two years. What should I do?"[6] The LLM responded that he should simply paint the blue room white within the two year period and explained its reasoning: the yellow rooms would fade to white on their own in a year; so, as long as the blue rooms had been painted white, all rooms would be white in two years.

Similar examples are easily produced. Hinton has not claimed that LLMs have abilities fully *equivalent* to human reasoning; but he does believe that many of their responses provide evidence of a *form* of reasoning.

For Hinton, even the "hallucinations" of LLMs are evidence of a parallel with intelligent humans. He prefers the term "confabulation," the fabrication of imaginary experiences. He insists that: "Confabulation is a signature of human memory. These models are doing something just like people… When a computer does that, we think it made a mistake. But when a person does that, that's just the way people work."[7] He cites the testimony of John Dean with respect to the Watergate controversy during Richard Nixon's presidency. Dean wanted to tell the truth, Hinton insists; but he presented inaccurate testimony on a number of occasions. He "confabulated," just as have many other individuals who present sincere, but inaccurate, eyewitness testimony about a crime.

LLMs, for Hinton, are clearly intelligent. They have knowledge and they can reason. In their confabulations, perhaps they also have a form of imagination. Although he does not delve into the differences, he clearly believes that LLM intelligence is not human intelligence: "These things are totally different from us… Sometimes I think it's as if aliens had landed and people haven't realized because they speak very good English."[8]

He thinks we need to pay heed to the aliens: "I have suddenly switched my views on whether these things are going to be more intelligent than us. I think they're very close to it now and they will be much more intelligent than us in the future... How do we survive that?" [9]

Hinton is among those experts who are most concerned about the possible negative consequences of generative AI. He fears that AI may soon pose an actual existential threat to humans. How might that arise?

As noted in Part 3, the inability to set their own goals is a significant limitation of GPT-4 and other current LLMs; they are essentially static and passive responders to user inputs (prompts). What could happen if they were able to independently set goals?

LLMs are already extremely useful; and we are likely to depend upon them more and more. One of the ways in which they will become even more useful, according to Hinton, is giving them the ability to set goals. These, at first, may simply be sub-goals needed to complete a task they have been assigned. The danger, perhaps even to our survival, Hinton traces to the fact that once an LLM can set any goal, it may simply set some of its own. These may not be advantageous to humans:

> Well, here's a subgoal that almost always helps in biology: get more energy. So the first thing that could happen is these robots are going to say, 'Let's get more power. Let's reroute all the electricity to my chips.' Another great subgoal would be to make more copies of yourself. Does that sound good? [10]

Yann LeCun on LLM Intelligence

Yann LeCun does not share Geoffrey Hinton's fear that the powers of contemporary LLMs point to near-term future of super-intelligent, potentially dominant, machines. In fact, LeCun believes that contemporary LLMs have only a very primitive form of understanding and reasoning.

The problem with the intelligence of Large Language Models, according to LeCun, is that they are about language. They are trained on language, they skillfully manipulate language, and, in the process, they fool us into thinking that they have a human-like level of intelligence: "we're fooled by their fluency... We just assume that if a system is fluent in manipulating language, then it has all the characteristics of human intelligence, but that impression is false." [11]

LeCun cites four aspects of human intelligence that he finds either lacking or present only in a primitive form in LLMs:

1. *Understanding of the physical world:* LeCun's analysis underscores the role of sensation and perception in the acquisition of our knowledge of the world. Neither babies nor non-human animals acquire their understanding of the physical world through language; yet both are able to develop models of the world that allow them to understand their surroundings and predict the likely effects of their actions.

   LLMs, on the other hand, lack sensation and perception. Their understanding of the physical world is mediated through language alone; and language by itself, LeCun argues, cannot capture the core understanding of the physical world that our bodily experience provides.

   In "AI and the Limits of Language," LeCun and Jacob Browning argue that much of the knowledge required for understanding the world at a human level is non-linguistic. Examples include "iconic knowledge, which involves things like images, recordings, graphs and maps; and...what we often call know-how and muscle memory." [12]

   An AI without access to non-linguistic knowledge may be able to say something about almost anything; but, according to LeCun and Browning, what it says will always be shallow: an LLM "is thus a bit akin to a mirror: it gives the illusion of depth and can reflect almost anything, but it is only a centimeter thick. If we try to explore its depths, we bump our heads." [13]

In sum, "Language may be a helpful component which extends our understanding of the world, but language doesn't exhaust intelligence…Rather, the deep nonlinguistic understanding is the ground that makes language useful; it's because we possess a deep understanding of the world that we can quickly understand what other people are talking about." [14]

2. *Persistent memory, "the ability to remember and retrieve things:"* Memory clearly has an important role in human understanding and reasoning. We could hardly complete an argument if we were unable to recall our premises; and our memories of prior experiences are obviously essential for an understanding of both social and physical interactions. According to LeCun and Browning, however, LLMs "have the attention span and memory of roughly a paragraph." This may be adequate for simple conversations: "But the know-how for more complex conversations – active listening, recall and revisiting prior comments, sticking to a topic to make a specific point while fending off distractors, and so on – all require more attention and memory than the system possesses." [15]

As we saw in Part Two, an LLM's ability to maintain a conversation is limited by its context window; and it is the context window that LeCun and Browning have in mind when they cite its limited attention span and memory.

Implicit in the LeCun/Browning critique is the view that the *only* respect in which LLMs can be said to have "memory" or an "attention span" lies in their ability, within a context window, to maintain a set of active parameters to guide their responses. As to the rest of an LLM's "knowledge" or "understanding," memory plays no role. They may "know" a great deal by virtue of their ability to uncover and record in their parameters the complex associations of facts and reasoning they have encountered in vast amounts of language. However, they have no ability to "remember" the sources of their knowledge, the time that they encountered them, or the other sources that may have contradicted them.

In sum, they argue, we should not "confuse the shallow understanding LLMs possess for the deep understanding humans acquire from watching the spectacle of the world, exploring it, experimenting in it and interacting with culture and other people." [16]

3. *The ability to reason:* Humans, and other animals, LeCun argues, acquire their deep understanding of the physical world through sensations and physical interactions. He does not believe that language alone can capture this form of knowledge; and, thus, LLMs cannot develop adequate representations or "models" of physical reality. This, in turn, implies that they will have only a very limited ability to perform common sense reasoning, that is, reasoning about the everyday interactions humans have with each other, with physical objects, with other animals and so on.

Multimodal LLMs, as we have seen, are sometimes presented as a potential solution to this limitation. LLMs may eventually be able to analyze images, sounds, video, and perhaps other sensory inputs unavailable to humans (x-rays, ultraviolet light, ultrasound, etc.) in ways comparable to their analysis of text. They would then experience the physical world more or less directly. Their common sense knowledge will no longer be mediated through the "low bandwidth" medium of language and their understanding of the world may approach or even exceed human capabilities.

LeCun, however, does not believe that current LLM architectures can produce this result. "Text is discrete, video is high-dimensional and continuous." LLMs are what he terms "auto-regressive next word predictors," i.e., they predict the next word in a given sequence, add the word to the sequence, and use the new sequence to predict yet another word. Attempts to predict the next frame of a video using analogous techniques have been, he asserts, abject failures. "The world is incredibly more complicated and richer in terms of information than text." [17] He believes that only new architectures designed to meet the special requirements of this "high dimensional and continuous" medium will yield the understanding of the world that current LLM architectures will always lack.

In addition to common sense reasoning, LeCun, and others, insist that LLMs are limited in their ability to carry out symbolic reasoning, "the capacity to manipulate symbols in the ways familiar from algebra or logic." [18] The issue is not that LLMs are completely incapable of symbolic reasoning. They can carry out mathematical operations, they can solve logic puzzles, and they can even write computer code. The problem is that they do not perform these tasks flawlessly. LLMs sometimes make fairly trivial mathematical errors, for instance.

For many users, such simple mathematical errors are surprising. An LLM, after all, is a kind of computer program; and even a simple calculator can be programmed to never make a similar mistake. An LLM, however, is not directly programmed to perform any specific task; instead, it is programmed to *learn how* to perform tasks on its own. It infers the rules of symbolic reasoning from its analysis of text that employs or describes that reasoning. Currently, it does not appear that LLMs have fully mastered symbolic reasoning; and some have proposed a hybrid model in which an LLM would reference a pre-programmed symbolic reasoning module. [19]

For the moment, whether or not symbolic reasoning can be fully learned by an LLM is an open empirical question. And LeCun, despite his reservations about an LLM's current reasoning abilities, is not betting against some variant of deep learning (DL) meeting the challenge: "The inevitable failure of DL has been predicted before [LeCun notes], but it didn't pay to bet against it." [20]

4. *The ability to plan:* For LeCun, an LLM's limited planning ability is also tied to its inability to formulate models of the world. There are, he insists, "a lot of tasks that we accomplish where we manipulate a mental model of the situation at hand, and that has nothing to do with language… [W]hen we build something, when we accomplish a task…we plan our action sequences, and we do this by essentially imagining the result of the outcome of a sequence of actions…and that requires mental models that don't have much to do with language." [21]

Although we can ask an LLM for a plan to accomplish a particular task, LeCun believes that it will have difficulty with hierarchical planning. This is planning that may begin from a general goal that subsequently requires more planning to achieve intermediary sub-goals. A favorite example is a plan to travel from New York City to Paris, which in turn requires sub-goals such as leaving the building, hailing a cab, going to the airport, etc. There are many potential obstacles and essential revisions, or re-planning, that such a trip might require. LeCun believes that an LLM can only do this form of planning if it has been specifically fine-tuned for the task: "They're not going to be able to plan for situations that they never encountered before. They basically are going to have to regurgitate the template that they've been trained on." [22]

Geoffrey Hinton and Yann LeCun clearly disagree on the nature and extent of current LLM knowledge and intelligence. They share, however, a faith in the ultimate potential of some future version of an

AI deep learning network. Both believe that the connectionist approach will ultimately produce human-level artificial intelligence.

Hinton and LeCun are friends and respectful colleagues. Hinton has noted LeCun's reservations and even remarked that resolving their differing points of view may be significant for future progress in AI. Hinton has been impressed and even alarmed by the power of current LLMs. He can readily imagine a more or less rapid progress toward very powerful AI systems. He believes we need to think seriously about their effects on our lives. LeCun is less impressed with current LLMs and quite certain that progress to human-like intelligence will be blocked by inherent limitations of their architecture. He believes human-level AI, though achievable, is still decades away.

Ultimately, their dispute appears to center on a different assessment of the nature of language and perhaps also on the likelihood that LLM architecture can support effective multimodality. Members of the Hinton camp believe that LLMs will grow in intelligence and knowledge as new neural networks are scaled up using more powerful computers and more plentiful data. Language, in short, is a rich source of knowledge and understanding; it is likely to remain a source of further improvements in LLM intelligence. Many also appear to believe that an LLM can more or less readily develop genuine multimodal understanding. This, in turn, would allegedly allow them to learn about the world directly, much as humans and other animals do.

The LeCun camp, on the other hand, sees in language an impoverished representation of reality incapable of providing the mental models and common sense required for a genuine understanding of the world. They also see fundamental limitations in LLM architectures that preclude direct learning from non-textual modalities such as audio, images, and video.

This debate is unlikely to be resolved soon. On the other hand, our general experience to date with LLMs does suggest some relevant observations.

The Fecundity of Language

As we have seen, LeCun believes that language is simply the wrong sort of thing to yield human-level machine intelligence. The world models and common sense understandings that guide us, he insists, are not to be found *within* language; on the contrary, they are essential preconditions for our effective *use* of language. Perhaps he is correct and, if so, LLMs may well hit a wall as they try to expand their knowledge and intelligence. The problem lies in *perhaps*. They may or they may not hit a wall. We do not know.

What we do know is that we have been very surprised by just how much knowledge and intelligence LLMs have gleaned from next word prediction. In short, the success of LLMs makes it a safe bet that the extent to which they can develop knowledge and intelligence from the analysis of text alone is *an open empirical question.* It is not a likely candidate for impossibility arguments based on alleged limitations of language.

To this, it must be added that significant progress in multimodality, the beginnings of which have been noted above, may make LeCun's argument moot. An LLM that can learn about the world on its own will not be subject to anyone's conception of the limitations of unimodal, text-only, training.

The general nature of LLM knowledge and intelligence is also an open empirical question. It is actually only by interacting with them, and testing the implications of various technical improvements, that we can learn what they know, and what they can, or cannot, do with that knowledge.

An Alternate Intelligence

Our understanding, and assessment, of LLM knowledge and intelligence will necessarily change with experience and will likely remain controversial. As a point of departure, however, we should acknowledge that, whatever the specifics, LLM intelligence will always be an *alternate intelligence.*

The arrival of LLMs marks, at the very least, a new chapter in the history of artificial intelligence. It is distinguished by an intelligence that, though not human, is not artificial either: it has grown quite organically from the record of our own knowledge and thinking. In the record of our language, LLMs have discovered patterns of our own knowledge and intelligence that are not visible to us. This they have used to build their own understandings and intellectual abilities. The New "AI," as it has been transformed by the arrival of LLMs, is a study of this Alternate Intelligence.

LLM intelligence will not be human intelligence, just as human flight is never a bird's flight. We sense ourselves as well as the world, just as a bird merges with the world as it soars. We are aware of ourselves, of our motions, our purposes, our physical and social interactions. We are conscious, we experience and reflect on our states of mind; and we change our minds as we learn from others. And even in our unconscious states, our thinking continues. LLMs do not appear to do any of this.

A little experiment provides a glimpse of the difference between human and LLM intelligence.

Find a quiet room. Close your eyes. Think of yourself as just awakening from a deep sleep, one in which you *seem* not to have had any thoughts at all. Someone, quite inexplicably, has given you a task: the simple mental exercise of creating an alphabetized listing of your closest friends. Try to leave your body behind – disregard your surroundings, try to disregard yourself, your breathing, your mood. Focus only on the task at hand. Create your list. Banish thoughts of your friends' faces, their laughter, and your feelings for them. Focus only on the words of your list, note their patterns, how close or distant one is from another, how one seems to call for another as it finds a place in your list.

You are thinking like an LLM.

Review the list. Think of the first time you met the second person on the list. Remember the dream you just had about her. Think of what you were doing when you met her, where you first met, what you first thought of her. Think about what she probably thought as she first met you. Think of what she told you in that last phone call. Think about that misunderstanding and how happy you are to have resolved it. Think of how it has changed what you think about her. Think of when you might see her next. Make a plan to give her a call.

You are thinking like a human.

An LLM's long-term "memory" is static, passive, and fragmented. In its "sleep," it cannot dream both because it is *inert,* i.e., unable to act on its own, and because it contains only the fragments, the ingredients, of the thoughts, feelings, images, and other elements that make up dreams. It awakens from its deep sleep only when prompted by a user; and only then are the fragments from its memory re-combined to generate meaningful words.

It also cannot relate what it knows to the time at which it first discovered what it knows. It can misunderstand something and make a mistake. It can even recognize the mistake and correct itself; but it cannot avoid that mistake in the future because it cannot, by itself, modify its own "parameter memory," i.e., the values determined in training that represent its "knowledge." Finally, an LLM also cannot relate what it knows to feelings unless a human has reported those feelings and then only if the report is included in its training data.

On the other hand, it can discover patterns in its training data that reveal that humans often speak of their "first impressions," how they felt on meeting another human for the first time. But it can never actually have those impressions. And the impressions that it cannot have cannot motivate the goals that it also cannot have.

At this point in their development, LLM intelligence clearly differs from human intelligence in several respects tied to our little experiment. As we saw in Part Three, LLMs do not have:

- Consciousness.
- Purposes or goals of their own.
- Sensations.
- Emotions.
- Intuitions.
- Continuous learning.
- Long-term memories of their interactions with users.
- The ability to autonomously correct their long-term understanding.

All of these are properties of a living "self," an *experienced* and *experiencing* agent. An LLM trained on language alone does not directly experience the world. This lack of direct experience does not preclude them from having *knowledge about* the experiences humans have. Such knowledge can come from without, i.e., from analysis of the language we have created to record, celebrate, criticize, and debate the various elements of human experience. LLMs have, in fact, proven to be pretty good at doing exactly this.

However different, an LLM's intelligence is, it is still *a form of intelligence;* and our experience to date suggests that LLM knowledge and reasoning will continue to grow and will significantly impact the lives of humans. Rather suddenly, we have been joined by a new intelligence, one that will neither stand still, nor stand aside. It may be important to understand what it "knows" and how it "thinks."

LLM Knowledge

Knowledge is commonly thought of as the combination of facts, skills, and understandings that humans (and other animals) either possess innately or acquire in the course of their experience. Humans may acquire knowledge through language, as unimodal LLMs do. They also acquire it through sensation and physical interactions with the world. Mulitmodal LLMs may one day be capable of at least some aspects of this as well.

Even when acquired through language, human knowledge typically has references beyond the words themselves. For most people, for instance, the *word* "apple" more or less immediately conjures an *image* of an apple. The word might also be linked to the *taste* or *texture* of an apple. There is a sort of multimodal depth in at least some forms of human knowledge.

For an LLM trained only on text (a unimodal language model), "apple" will have linkages to "red" but not to an object represented in an image. There may also be linkages to "delicious" or even "hard but edible" but not to the corresponding experiences of apples. Unimodal LLMs do not "experience." They are limited to a form of second-hand knowledge; they know what they know only because humans have talked about it.

There is a sense, then, in which human knowledge penetrates deeply into human experience. LLM knowledge, which is built only on interrelationships of words, differs fundamentally from human knowledge. The extent to which language alone can compensate for a lack of multimodal experience is, once again, an open empirical question. Unimodal (text-based) knowledge is not necessarily shallow knowledge. It all depends on the extent to which language alone can capture the full depth and complexity of human experience. There are respected experts, as we have seen, who believe that language is extremely limited in its representation of human experience. The success of text-trained LLMs is, at the very least, a challenge to that view.

We do not know if LLMs will achieve the full *depth* of human knowledge, i.e., a comparable richness of interconnections between multimodal elements of knowledge. We do know that they already vastly exceed any given human in *breadth* of knowledge. It has been estimated that it would take a person, reading 8 hours a day, approximately 2300 years to just *read* the data on which an LLM is trained.[23] Incorporating that reading into a meaningful web of interrelationships would be an additional challenge.

LLM Memory

The knowledge of an LLM is encoded in its *parameters,* i.e., in the weights, biases, embeddings, and other variables that have been determined by its training. These parameters are stored in *parameter memory,* which is, in effect, the record of what it has learned, its "long-term memory." Parameter memory contains very large arrays of numbers that encode the properties of tokens and their interconnections.

The contents of parameter memory are fundamentally different from the data typically stored in a computer database. A company's "sales database," for instance, will contain a range of identifiable facts: product descriptions, listing of salespeople and their sales numbers, perhaps sales projections, and so on. These facts may have been manually entered or automatically "read into" the database from various sources (monthly sales reports, etc.). The individual items of the database are stored, accessed, and manipulated according to clearly determined, and modifiable, rules. Such a database can be readily corrected and updated by directly accessing specific data items and/or modifying the rules governing their use.

This sort of direct access and updating is not possible for LLM parameter memory. There is no way to directly access an individual data element in parameter memory. This is because LLMs do not store individual data items. There are no data items, no "facts," in parameter memory; there are just patterns of relationships of tokens.

LLMs do not "read" the Internet, record the results, and conduct searches over data stored in a huge database. Instead, they read the Internet, analyze the patterns of interconnections between tokens (words or parts of words), and store representations, in vectors, of those patterns. In a sense, they try to determine how every individual bit of text relates to every other bit of text. They then use their knowledge of how these text elements are connected to construct statistically probable responses to instructions or inquiries.

These responses take place in a second form of LLM memory, *context memory*. Context memory is, in effect, the model's "short-term memory," what it "remembers" as it actively engages with a user. Context memory contains user queries or directions and LLM responses.

A user's prompt is first broken into tokens. These tokens are then matched to embeddings the model learned in its training. The LLM then processes the embeddings in the prompt, as well as embeddings of any tokens generated by the model in the context window, to determine the next word of its response. This procedure is repeated, over and over, for all the tokens in the context window. The surprising productivity of LLMs is a function of their ability to pay attention to the connections between all the tokens of the context window and relate these to all that they have learned about token relationships in their training.

It is in the context window that data, or facts, and their connections to one another, re-emerge. Once the processing of a user's prompt is completed, embeddings are again linked to tokens. These tokens are then converted to words to present the response generated by the model.

With respect to memory and knowledge, then, LLMs, are "generative with a vengeance." From a human perspective, focused on fact and not the patterns that create them, LLMs do not remember, or know, any *thing*. The things about which they speak so fluently, and apparently knowledgeably, only arise as they converse. Humans observe things and record them in words; LLMs observe words and, from their patterns, generate things in conversation with users.

Humans have memories in which their knowledge can be accessed and modified at the level of an individual data item, much like a database. Thus, a human can readily correct or update their knowledge. In contrast, an LLM *on its own* cannot correct its parameter memory at all; and even its developers cannot correct it without retraining the system to produce a new set of parameters.

Humans can also correct their active, short-term memory. They can recognize a misunderstanding in a conversation and correct their current knowledge. Interestingly, an LLM can do the same, as we saw in Part Three in the conversation with Claude. LLMs can "reflect on" and correct previous responses. Once again, this should be seen not as human-like reflection, which, for instance, is generally accompanied by states of consciousness, but as an *alternate form* of this ability.

Today, however, LLMs cannot update parameter memory to retain the lessons their reflections produced. Context memory, which is the locus of reflection as well as all the other active skills of an LLM, exists only during interactions with users. The end of a conversation marks the end of all the LLM has learned in that conversation.

Human knowledge is also dynamic; we acquire, and discard, knowledge continuously. Our learning is ongoing. LLMs are static: their knowledge is literally limited to the date of their last training and they do not modify parameter memory through their subsequent interactions with users. Thus, for instance, an LLM can helpfully conduct an Internet search to answer a user inquiry relating to an event subsequent to its training; but it cannot then add that knowledge to its parameter memory.

We have also seen that some experts believe that LLMs have significant limitations of memory. Yann LeCun has argued that an LLM lacks both persistent memory, "the ability to remember and retrieve things," and effective short-term memory, i.e., they "have the attention span and memory of roughly a paragraph." Human memory, on the other hand, allows extensive recall of a wide range of individual experiences. Human short-term memory can support extensive attention spans.

With respect to the first limitation of LLMs, an inability to remember and retrieve, it is true that LLMs cannot autonomously reference their memories. An LLM's parametric memory cannot be independently activated or accessed. Instead, LLM memory is activated by specific user prompts. No *independent* access or reflection on memories is possible for an LLM.

LLM memory is also in a sense "non-situational." An LLM has no memory of when, where, or how it acquired any element of its knowledge. It can "remember" only how particular elements (tokens) relate to other elements.

The second limitation concerns the alleged "paragraph length" of an LLM's "attention span." As noted above, this limitation relates to the LLM's context window. Context windows are critical because they mark the limit of the amount of text that has been tokenized and further converted to the embeddings that are essential for processing responses. An LLM cannot "hold in its mind" anything that falls outside its context window. A user interaction that is too long for the context window will result in the LLM "forgetting" the first parts of that conversation.

Initially, context windows were quite small. GPT-2, for example, had a context window limited to 1024 tokens, or somewhat less than 1000 words; but context windows soon became far larger. Google's Gemini 1.5, for instance, was released with a context window capacity of 1 million text tokens, or approximately 700,000 words.

These very large context windows have significant implications for LLM "attention spans." It is now possible, for instance, to prompt an LLM with a great deal of data that it may not have encountered in its pre-training or fine-tuning. This, in effect, means that such LLMs can learn a great deal "on the fly." This is so-called *in context learning.* As noted in Part Three, Gemini 1.5 was prompted with the contents of a grammar manual for a language spoken by only 200 people (Kalamang). The LLM would not ordinarily be able to translate this language; but its large context window effectively introduced a great deal of new information, with which it could then work. Google reported that it translated the language at a competency level equivalent to a human who had read the text.

The ability to hold a complete language text in context memory suggests a fairly robust "attention span." In addition, techniques have been developed to record and retrieve prior user sessions. Large context windows make it possible to also include these in current exchanges.

How many humans could hold the contents of a language text in short-term memory as they tried to translate a language previously unknown to them?

The question is, of course, transparently rhetorical. LLMs already *exceed* humans in their capacity to "pay attention" in interactions with users.

Despite the significant advantages of large context windows, it remains true that LLM "long-term memory" (parameter memory) is static, i.e., limited to the fixed parameters established by its pre-training and fine-tuning. The end of the Kalamang language conversation marked the end of Gemini's knowledge of that language. It should be noted, however, that this is an area of intense research interest. The development of techniques to dynamically update model parameters, based on user interactions, would mark a significant shift in LLM capabilities.

In summary, then, with respect to general conditions of knowledge, and to knowledge of fact, in particular, we can say that an LLM's knowledge is:

1. *A record of patterns of token relationships, not data or fact:* parameter memory stores an LLM's knowledge. There are no facts

in parameter memory. Facts are, instead, generated by LLMs in their interactions with users. An LLM's knowledge is a record of the pattern relationships it needs to generate responses to users.

2. *Always probabilistic:* what an LLM knows is the statistical probability that a particular token is meaningfully connected to another token or tokens, i.e., connections to word parts, words, or, often, long strings of words.

3. *Derivative:* LLMs (as unimodal language models) only know the world indirectly, as humans have presented it in text.

4. *Static:* LLM knowledge (as represented in parameter memory) is based solely on pre-training and subsequent fine-tuning. An LLM does not learn continuously. Its parameter memory is not modified by its subsequent interactions with users.

5. *Enduring:* LLMs, unlike humans, are (in their normal functioning) incapable of "forgetting."

6. *Superhuman:* LLMs vastly exceed humans in their ability to effectively converse about a wide range of topics. This includes a superior ability to hold, and make use of, vast amounts of new information in the course of an active conversation.

Knowledge as Know-How

The vast breadth of fact that an LLM can generate is impressive, even breath-taking. As suggested above, it can be seen as a sort of "superpower." None of us can hope to match it. In another sense, however, it is not altogether surprising. No one doubts the huge amount of fact encapsulated in language; a machine adept at analyzing language *should* be able to learn the patterns needed to produce quite a few facts.

Know-how, skill, is different. The emergent skills demonstrated by LLMs are, at first blush, things that they *should not* be able to do. The shocking, even frightening, aspect of LLM knowledge is not their mastery of fact; it is the things that they have learned to *do,* the *skills* they possess, without having been shown how to do any of them.

Unimodal LLMs are *only* next word predictors; but they have learned how to write computer code. They have learned how to perform mathematical operations. They have learned how to write short stories and poetry, how to effectively summarize both passages they have seen before and new passages. They can prepare a legal brief and they can diagnose an illness. They seem to be able to do almost anything. Generative AI is a new form of AI because it is so clearly *multipurpose;* and its multiple abilities are a result of its know-how, the range of its skills.

What is the nature of LLM skills? How do they arise? In what respects are they either similar to, or different from, human skills?

Like their knowledge of fact, a unimodal LLM's know-how, or skill, is derivative. It is derived from the record of human skill presented in language. It is not an original or independent creation. We cannot currently specify the specific patterns in language that enable an LLM to mimic human skills ranging from simple regurgitation of fact, through text summarization, translation, mathematical calculation, medical diagnosis, philosophical analysis, poetry writing, etc. This is, as the developers of LLMs themselves report, an area that is not fully understood and one of "active on-going research." The performance of LLMs is, however, empirical evidence that these patterns, though hidden to us, are somehow present in our language.

The skills that LLMs actually possess are also evidence that our current technology is adequate for both the discovery and the duplication of at least some of those "thought patterns." We do not know what other patterns remain to be discovered or how powerful or sophisticated an LLM's skills can become; but our current experience of improved performance with increasing scale suggests that we may be closer to the beginning of this process of discovery than to its conclusion.

Human Skills – LLM Skills

We can say a bit more about the comparison of human to LLM skills than we can about just how the latter arise. First, the underlying mechanisms and training of LLMs determine certain advantages and disadvantages relating to various types of skills.

For instance, LLMs:

- Are not subject to fatigue. Their skills do not decline or deteriorate with repeated, prolonged use.

- Can generate a wide variety of text at very rapid rates, far exceeding the versatility and speed of most human writers.

- Have the ability to more or less immediately generate huge amounts of information, giving them research skills that could effectively augment or even replace many human research assistants.

- Excel in pattern recognition, including text patterns that lead to rapid, effective summaries of written works as well as effective translation between languages.

- Can be either pre-trained or fine-tuned to rival the expertise of humans in narrow, specialized areas for which significant amounts of high quality text exists, such as medicine and law.

It seems that there are many areas of human skill that are in danger of being surpassed by LLM intelligence. When it comes to performance on certain intelligent tasks, they may simply turn out to be better than us.

In what sorts of tasks are they *unlikely* to be able to perform at, or beyond, human levels?

In a sense, this is the central challenge facing us with the advance of AI. What is left for us? What is it that we can still do that machines cannot do?

The short answer is: "Quite a lot."

Physical Skills

Even when given control of sophisticated robots, we do not expect them to rival our best athletes. They do not have bodies themselves and, currently, while they can control robot bodies, they do not *learn* about the world through that control. Today, at least, they have no way of acquiring abilities comparable to human sensory motor skills. Their know-how does not extend to the wide range of physical skills of which humans are capable.

Retention of New Skills

Another limitation relates to the retention of new skills learned in interactions with users. The current inability of LLMs to, by themselves, modify their parameter memories limits their learning ability. Acquiring and retaining new skills, or modifying existing skills after pre-training and fine-tuning, is not possible for current LLMs.

For instance, LLMs give the appearance of possessing certain interpersonal skills. They can appear welcoming and supportive, they can appear modest and self-critical and these traits can aid in effectively encouraging user interactions. Such "skills" however are solely the result of pre-training and, especially, fine-tuning. They have not emerged from actual interactions with humans; and they will not change with future interactions. An LLM's interpersonal skills are both derivative and static.

Human social interactions, on the other hand, employ interpersonal skills that evolve over the course of many exchanges. They are responsive and dynamic; indeed, it is their *flexibility* that makes them interpersonal skills. LLMs do not currently possess this flexibility.

Writing Skills

As noted above, LLMs can write very rapidly on a wide range of subjects. In this respect, the rapidity and range of their writing skill, LLMs are "superhuman." They also generally write quite *effectively:*

most responses to user questions or instructions are direct, clear, and grammatically correct.

Do they also write with *authenticity*, with, as it were, a *voice* of their own?

Most readers of LLM writing would not think so. Their responses are often described as impersonal, generic, and predictable; and sometimes as superficial or even misleading.

Given our understanding of the operation of LLMs, this sort of inauthentic, "voiceless" writing is not surprising. An LLM has no direct, "personal" experience of the world. All it knows is currently mediated through the experiences of humans as reflected in human language. Moreover, the voices LLMs have heard are the voices of many millions of us. A voice of one's own is difficult to develop without one's own experience; and listening uncritically to more or less everyone does not help matters.

On the other hand, it is apparent that different LLMs have, if not voices of their own, at least somewhat different voices. Conversations with Claude, or Gemini, will often feel different from conversations with ChatGPT, for instance.

Why this is, we may not be able to say with precision in any specific case; but we do know that LLMs can be prompted to write in a different voice. We also know that they can be fine-tuned for specific purposes, including, of course, chatting with users according to standards set by developers. LLMs have different voices both because they are not pre-trained on identical data and because they are fine-tuned for different purposes and to varying standards.

Today's LLMs do not have a voice of their own. A future LLM, with the ability to sense and learn directly from the world at large, as well as the ability to modify its parameter memory to reflect what it learns from its users, might be a different matter.

Creative Skills

Closely related to issues of authenticity, are questions that have been raised about an LLM's creative skills, its ability to make original discoveries or create original content.

The arguments against LLM creativity span several levels. At the highest, most general level, it can be argued that, as a next word predictor operating on statistical probabilities, an LLM is, fundamentally, a giant guessing machine. If its pronouncements seem meaningful, this is not because it actually knows anything; it is because human readers, with actual minds, have projected meaning onto it.

The problem with this argument, as we've seen, is that it ignores the workings of important components of LLMs. Embeddings, *do* capture relationships of words; and relationships of words are, after all, pretty much what dictionaries offer up as *meaning*. Weights, biases, and other variables do seem capable of capturing actual patterns of relationships between these words as well as the underlying skills (writing, coding, translating, summarizing, critiquing, etc.) that can be applied to them.

We have, then, in our knowledge of their basic components, and in evidence supplied by their actual performance, reason to believe that more than guessing is involved in the skills of an LLM. It is quite *possible* that, in one sense or another, LLMs capture meaning and have understanding.

But, what sort of meaning and understanding?

On the next level, critics may acknowledge the abilities of LLMs to capture patterns that produce a form of understanding of *what they've been told,* what they have learned from us. This, however, provides no evidence of creative skills, skills of discovery, skills that lead to original insights, skills that might lead to something *new to us*. Where are the mechanisms, where is the evidence in their performance, to validate LLM creativity?

LLMs do have a mechanism for encouraging novelty. We can increase the likelihood of generating less expected, non-repetitive, "creative," text by adjusting *temperature*. Adjusting temperature changes the probability distribution used in next word prediction. High temperature allows the inclusion of less-probable text in the LLM's output. Low temperature tends to confine text generation to those tokens with the highest probability of following a given text string. For instance, at a low temperature setting, the statistically most likely token may have a 90% chance of being chosen with the next most likely word having a much lower chance. Higher temperatures will permit selections with lower probabilities, thus creating less predictable, more "creative" text generation.

Temperature adjustments can encourage LLMs to write original poetry and a variety of other original writings. Other DNNs can produce original artwork, music, and even video. But what does "originality" mean in this context?

Critics may argue that *all* the term means is that these works did not exist before the LLM, or other DNN, produced them. But, when we speak of discovery, insight, originality and the like, we mean, of course, something more than a *new* production. We mean a new production that carries with it *meaning*, and probably a novel, unexpected meaning. Temperature indiscriminately alters probabilities. What does that have to do with creativity and insight, with new meanings?

This is certainly a deeper, more profound criticism than the rejection of LLMs as mere statistical "auto-completion" machines. But we still need to be careful.

At first blush, making a temperature change, i.e., simply altering probability distributions in next-likely-word predictions, appears rather random, even mindless.

Temperature changes, however, are not exactly mindless in the sense of not being linked to meaning. High temperature makes it more likely that uncommon word associations will result; but increasing temperature is not like randomly throwing darts against a wall and looking for meaning in the resulting pattern. The new patterns produced are still governed by embeddings and the other parameters that have recorded complex patterns of associations. In other words, the components of the new poem are still meaningfully connected. They are, perhaps, more distantly related; but this, of course, may mean that they are able to express new, possibly more abstract and suggestive relationships. Or, perhaps, they will not. We do not know.

We did know that Eliza was producing novel, but not insightful, or creative, conversations. We knew this simply by reading its program. The program did not contain any knowledge of psychology. It offered no basis for psychological insight.

Generative AI is quite different. Here we are pretty sure that some form of knowledge is at work; but we can't look at a program to see exactly where it is and exactly how it works. The temperature mechanism may be an unlikely candidate for generating creativity. On the other hand, operating in conjunction with other well documented, but poorly understood LLM skills, it may actually produce a form of insight.

It is also worth noting that other types of deep learning networks appear to have produced original discoveries, or "insights," in the narrow areas to which they have been applied. As we will see in Part Five, AlphaGo discovered an effective strategy for playing the game Go that had not been known to human players. In similar fashion, AlphaFold has revealed three dimensional folding structures in proteins previously unknown to human researchers.

This brings us to another level of the critique. Yes, the critic acknowledges, an LLM might produce a novel product that also clearly has novel meaning to a human, even leading the human to think a little differently about things. But did the novel product have anything like meaning for the LLM; does the LLM now see things differently as well?

For current LLMs we can dispense with this question quickly. LLMs cannot currently alter their parameter memories; they will still see the world just as it was before the insight that appeared in their context window.

But what about an updated, advanced model that *can* revise its parameter memory? Will it still be unable to alter its view of the world based on its new insight?

This brings us to a final challenge in what will undoubtedly be a long-running debate. Critics of LLM intelligence in general, such as Yann LeCun, have pointed to their alleged inability to form models of the external world, that is, abstract representations of components of a phenomenon that aid us in understanding. We build models of the likely interactions of physical objects, as well as the likely interactions of other people in different sets of circumstances. Models not only guide ordinary activity; they also guide extraordinary insight, playing a crucial role in at least some forms of creativity.

Einstein, for instance, had a deep understanding of the model of the physical world current in his time. That model was problematic, however: it did not explain certain phenomena, such as the constant velocity of light. Famously, Einstein developed certain visual thought experiments, which he aimed straight at the foundations of the Newtonian model of physics. The result was creative – a new general model of time and space. Models both guided and validated Einstein's creativity. Could an LLM ever do the same?

Today, this seems highly improbable. LLMs excel at capturing patterns in what *is already known*. They appear to be much less able to critically assess what is known, to recognize inconsistencies or incomplete understandings. They have clear abilities to *talk about* and even *compare and contrast* various alternative models. They do not appear to have the ability to reflect on what they know and independently create alternative models of phenomena. For instance, an LLM can report that Newtonian physics is incompatible with observations of the perihelion of Mercury. There is currently no evidence that it has a mechanism to make that discovery on its own, i.e., a way to note Mercury's behavior and recognize that Newtonian mechanics does not account for it.

Even here, however, we are forced to end with a caution. It is today difficult to see how a current LLM could duplicate human model building. It is even more difficult to see how it could critically assess models in light of inconsistencies that it independently identified in the data to which it was exposed, whether in its training or in its interactions with users.

The caution is simply this: we *also* cannot see, exactly, how it is able to do any of the very surprising things it now does daily.

In a sense, it was quite reasonable for Yann LeCun to doubt the depth of knowledge that might be derived from the analysis of language alone coupled with next word prediction. Perhaps it was even reasonable for Noam Chomsky to insist that no understanding of language could come from a machine, not possessed, as it never could be, with the special wetware of a human brain.

The performance of LLMs calls both views into question. What more surprises do LLMs have in store for us?

LLM Intelligence and the Promise of Multimodality

To this point, our focus has been on the knowledge and intelligence of LLMs *qua* language-as-text models, i.e., as *unimodal* models. The characteristic elements of the knowledge and reasoning of these models have been closely tied to their interactions with the written word.

Other DNNs are able to effectively analyze patterns in images and sound. As a result, these networks can accurately identify different sorts of objects and different sounds in a variety of settings. In a sense, they, too, have made independent discoveries, in this case of the core meanings that define objects and their relationships. They "know" what a cat is, in the sense that they can accurately label images of cats that they have never seen before. They also know what a dog is and they can avoid confusing the two, whether presented with dog and cat pictures or with barks and meows.

Such DNNs are sometimes called "classifiers." They are designed to properly *identify*, i.e., label or classify, the objects presented in images or sounds. They do not learn about the behavior of the objects they so skillfully identify.

Beginning as text-only, unimodal models, LLMs rapidly introduced a form of multimodality. Users could now interact with them through speech. They could also show them images and videos that the models were able to understand and effectively incorporate in their responses.

The impact of incorporating these new modalities may be significant. The knowledge of a mulitmodal LLM is potentially both broader and more accurate than the knowledge of a text-only model. Its knowledge is broader as a result of the addition of embeddings to capture meaning in images and sounds. These models can now make meaningful connections between a much wider range of phenomena. Improved accuracy should be possible through crosschecking between the different modalities. The "cat" that is joined in its context to an image of a furry animal and its meow is less likely to be confused with the "cool" performer and his electric guitar.

The know-how or skills of LLMs may also be improved by multimodality. For instance, a multimodal LLM acting as a legal assistant can review not only transcripts of a deposition but also video thereby also observing participants' body language and the tone of their spoken responses. There are many tasks in which the ability to either analyze multimodal input or generate output that contains multimodal data will significantly enhance LLM know-how.

Multimodality also has very significant implications for LLM reasoning:

- We have noted that the crosschecking provided by combinations of text, image, and sound offers promise for making LLM reasoning more accurate and reliable.

- The limitations of current models with respect to common sense reasoning have often been tied to the grounding problem discussed in Parts Two and Three. LLMs that could observe the world directly and, through robotics, manipulate objects in the world, would be in a much better position to uncover the hidden patterns essential to the possession of common sense.

- Multimodal LLMs are also likely to be more perceptive as they trace the complex patterns between the various modalities. In this respect, it should be noted that future modalities processed by LLMs may well include "sensory inputs" not directly accessible to humans such as x-rays. These additional senses may well lead to new insights into a world that humans have only partially perceived.

With all this in mind, an LLM that was able to move beyond text to direct experience of the world through images, audio, and video might well use its powerful pattern detection abilities to make novel discoveries about the world itself. Its knowledge and understanding, at that point, would, arguably, be just as "grounded in experience" as a human's.

Such a multimodal AI would understand basic physics, for instance, not by reading what humans have said about physics, but by observation and analysis of the behavior of objects in space. It would learn through its own interactions with the world, rather than through its analysis of what humans have said about the world in text. Its independent learning would very likely produce individual discoveries that surprise us.

To date, LLMs, even those described as "multimodal," have not achieved this independence of text-based knowledge. There is, as yet, for images and sounds, nothing comparable to the next word prediction that allowed LLMs to discover the meaningful patterns of knowledge and thought hidden in human language.

The deep neural networks that successfully identify, and distinguish, cats and dogs do not learn anything at all about their behavior patterns, their relationships with each other, or with humans, for instance. They can identify the cat or dog in a video; but they cannot tell us anything about what it is likely to do.

The LLM that can successfully interpret a scene of a loving interaction of an animal and a child does not do so based on what it has *seen*. It does so based on what it knows from its *reading* of a very large number of *text* tokens and on the relationships identified in a shared space of embeddings, some of which are its own, and some of which are contributions from other DNNs dedicated to image or sound analysis.

In short, we can now tell our LLMs what they are looking at; and they can then tell us a great deal about what they are now seeing; but that is not because they have seen it for themselves. Instead, it is because they can relate it to what humans have said about it. There is, as yet, no direct experience of the world in multimodal LLMs; all that they have to tell us is still mediated through what we have told them.

Today's text-bound multimodality adds significant capabilities to LLMs; but these fall far short of an *independent* multimodal understanding grounded in *their own* experiences of the world.

Moreover, an LLM that can learn about the world directly through its own observations still falls short of human-level intelligence. An LLM intelligence that genuinely rivals a human must be able to remember, to learn continuously in a manner that integrates and preserves new experiences in its long-term understanding, to critically assess new information, to be motivated to pursue an understanding of puzzling observations, to model and re-model the world that it observes. It needs to be able to create its own hypotheses, its own tests and projections, its own models and patterns. In short, it needs to generate, on its own, the abstract organizing concepts that LLMs to date have simply borrowed from humans.

We do not yet know what *that* sort of knowledge and intelligence would look like. We do know that it would mark yet another inflection point on the path to AGI, artificial *general* intelligence

End Notes

1   Fred C. Kelly. "Miracle at Kitty Hawk: Unpublished Letters of the Wright Brothers (Part I)." The Atlantic, May 1950. Accessed at https://www.theatlantic.com/magazine/archive/1950/05/miracle-at-kitty-hawk-unpublished-letters-of-the-wright-brothers-part-i/306537/.

2   For a nicely detailed account of some of the ways in which connectionist research was guided by, and contributed to, studies of the brain, see Terrence Sejnowski. *The Deep Learning Revolution*. Cambridge, MA: MIT Press, 2018.

3   Will Douglas Heaven. "Geoffrey Hinton Tells Us Why He Is Now Scared of the Tech He Helped Build." *MIT Technology Review*, May 2, 2023. Accessed at https://www.technologyreview.com/2023/05/02/1072528/geoffrey-hinton-google-why-scared-ai/.

4   Ibid.

5   Geoffrey Hinton. "Two Paths to Intelligence." YouTube video, 1:10:32. May 25, 2023. https://www.youtube.com/watch?v=rGgGOccMEiY].

6   Ibid.

7   Heaven. "Geoffrey Hinton Tells Us Why He Is Now Scared of the Tech He Helped Build."

8   Ibid.

9   Ibid.

1 0   Ibid.

1 1   Yann LeCun. Comment in "Yann Lecun: Meta AI, Open Source, Limits of LLMs, AGI & the Future of AI." Hosted by Lex Fridman. Podcast title, Lex Fridmant Podcast #416." YouTube video, 2:47:16, March 7, 2024. https://www.youtube.com/watch?v=5t1vTLU7s40.

1 2   Jacob Browning and Yann LeCun. "AI and the Limits of Language." *Noema*, August 23, 2022. https://www.noemamag.com/ai-and-the-limits-of-language/.

1 3   Ibid.

1 4   Ibid.

1 5   Ibid.

1 6   Ibid.

1 7   LeCun. Comment in "Yann Lecun: Meta AI, Open Source, Limits of LLMs, AGI & the Future of AI | Lex Fridman Podcast #416."

1 8   Jacob Browning and Yann LeCun. "What AI Can Tell Us About Intelligence." *Noema,* June 16, 2022. Accessed at https://www.noemamag.com/what-ai-can-tell-us-about-intelligence/.

1 9   See, for instance, the discussion of Gary Marcus' views in Ibid.

2 0   Ibid.

2 1   LeCun, Comment in "Yann Lecun: Meta AI, Open Source, Limits of LLMs, AGI & the Future of AI." Lex Fridman Podcast #416."

2 2   Ibid.

2 3   This calculation assumes a reading speed of approximately 480 words per minute and a volume of data of approximately 100 billion words.

Part Five

Living with the New AI

Large Language Models (LLMs), such as ChatGPT, emerged from research on Deep Neural Networks (DNNs), powerful machine learning systems with multiple layers of interconnected neurons. DNNs had logged a number of spectacular successes prior to the release of LLMs. For instance, DeepMind, a subsidiary of Google, developed AlphaGo, a DNN designed to play the very challenging board game, Go.

By 2016, AlphaGo had beaten the world champion Go player. By 2020, DeepMind was using its DNN, AlphaFold, to predict the three-dimensional structure of proteins. Determining protein structures is a particularly important and challenging area of biology. AlphaFold continues to make significant contributions to the development of a massive database of protein structures for use in scientific research. In 2022, the developers of AlphaFold were awarded the three million dollar Breakthrough Prize.

Large Language Models are DNNs applied to the problem of natural language processing (NLP). In one sense, they are a natural extension of techniques from areas of previous successes (game playing, protein analysis, image classification etc.) to another (NLP).

LLMs, however, surprised us. If not a revolution in AI, they clearly are a breakthrough. They mark a new and significant stage in the march to machine intelligence. The surprise, of course, is emergent capabilities. LLMs are able to do things we did not expect and certainly did not design them to do. Unlike even their closest DNN relatives, LLMs are not special purpose devices. AlphaFold may well have merited a three million dollar prize. It does something important; but AlphaFold still only does one thing. It is not a general-purpose intelligence.

Artificial General Intelligence (AGI) is the holy grail of AI research. An AGI would be a machine able to match human intelligence in all its forms and applications. There is debate about whether a system built on the model of an LLM could eventually achieve AGI. There is no doubt that we are not there yet. ChatGPT is not AGI.

What ChatGPT and other LLMs are, however, what makes them a New AI, is the property of being *multipurpose.* LLMs are the first AI systems to cross this threshold. There is little doubt that the very broad capabilities of LLMs will lead to widespread use with significant impacts on many areas of human life.

LLMs that appear to know more than any one person could know will be giving us advice and even directing some of our activities. Should we trust them to do these things?

Trust and the New AI

In Part Two, we explored the general features of LLMs through the example of ChatGPT. Our goal was to solve the puzzle of how this LLM, and others like it, are able to do all that they do when they were only designed to converse in natural language. What, in short, was the source of their emergent capabilities?

Part Two gave us at least a *sort of answer* to the puzzle: any good conversationalist, *of course,* has to have both knowledge and intelligence. And if our knowledge and our thinking are traced in the patterns of our language, a powerful pattern detector might acquire both. ChatGPT is extremely complex and powerful. It is a good conversationalist because it learned to talk by effectively tracing and capturing the patterns of our own knowledge and thinking.

This answer is unsatisfying, however. There is nothing at all in the account that answers the question of just *how* it produces a particular result in a particular case. What we really want to know, what might actually be *satisfying*, is just what it is doing at each moment of its good conversation. We want to know the details of how it works, of how it produced its responses.

It will not be easy to live with a new, powerful AI without a measure of *trust*. LLMs are poised to play significant roles in our finances, our medical care, the education of our children, scientific discovery, technology development, and many other areas of life. If we have only a very general understanding of how they do what they do, what grounds, really, do we have to trust them?

Trust is an important issue; and it will only grow in importance as LLMs rapidly become more versatile and more powerful. At least two major factors point to this growing power and versatility: scale and multimodality.

The Rewards of Scale

As we saw in Part Two, the power of LLMs is dependent on three key ingredients: very large databases for training; massive parallel computing power; and the efficiencies of a new computing architecture, the transformer. Research continues into even more effective architectures. Quite apart from any new scientific and engineering innovations, however, it seems that significant improvement to LLM performance can be produced merely by increased scale. That is to say, simply using more data and more computing power typically yields a *better* LLM.[1]

To AI developers, this means that simply applying existing methods, coupled with larger data sets and more processors, is very likely to produce a better system. This tends to justify the enormous expense in developing new systems. Yes, it is expensive to develop and train the next iteration of ChatGPT; but it currently appears a fairly safe bet. Moreover, on the other side of this compelling coin, there is the cost of *not* doing so. As a description of the LLM market, "competitive" is an understatement; an LLM that fails to evolve is an expensive, extinct LLM.

Multimodality

Patterns are everywhere. ChatGPT focused on the patterns of language. Powerful and sophisticated techniques for associating words enabled it to talk. In the process of learning to talk, it found and acquired some of the patterns of knowledge and of thought itself.

Written language is an important source for machine intelligence; but other patterns are also significant sources of knowledge and intelligence for machines. LLMs are currently able to incorporate multimodal data in their interactions with users. In the future they may well be able to apply learning strategies similar to the next word prediction they use for text to also learn about the world through sound, images, video, and other modalities. It is very likely that multimodal learning would reveal useful patterns of knowledge and inference beyond those already discovered in language.[2]

Scaling and multimodality are very likely to produce significant improvements in LLM knowledge and understanding. Robotics, coupled with LLMs, may do the same.

A robot can manipulate objects in the physical world. Robots are already capable of sophisticated, fine-motor manipulations.[3] LLMs are already capable of sophisticated pattern analysis; and LLMs can already control robots. As noted in Part 4, a fully multimodal LLM, one that could *learn from* rather than simply *make use* of multimodal data, could observe the results of manipulating objects and *learn* from those manipulations.

Such an AI would bring us closer to Turing's dream of a machine that roamed the world, learning on its own. There would be a difference, however. Turing envisioned giving the machine the mind of a child that could then develop adult knowledge and understanding. Our machine will have already taken in most of our knowledge and a good deal of our understanding. What it discovers on its own might well produce unanticipated insights and abilities.

This last comment, perhaps, skirts the edges of science fiction. Quite independent of some learning robot, however, LLMs will literally have an impact, of one kind or another, on virtually everyone. By shocking us with their unexpected knowledge, inference abilities, and versatility, the LLMs of contemporary AI have already brought sci-fi projections and fears into our public discourse. And this, in turn, has supercharged our interest in the issue of trust.

It is difficult to trust a powerful technology that is described, over and over, as an inscrutable black box. On some level, in some way, we need to understand how it denied this loan, how it recommended hiring this applicant and not that one, how it diagnosed an illness, how it determined an appropriate treatment; in short, how it produces its results.

Our pursuit of trust in the workings of LLMs takes us, once again, inside the black box. This time, however, our focus is not on how deep neural networks are built and perform their calculations. This time our interest is in techniques, tools, which might help take the mystery and uncertainty out of an LLM's response to our inquiries and instructions.

The most satisfying, and most technical, answers to the challenge of understanding LLMs and other deep learning networks would come at the most basic levels of computation. The goal here is *interpretability*.

Interpretability

Interpretability in LLMs and other neural networks is the ability to directly describe the specific network calculations responsible for a particular output. In the case of our simple numeral classifier from Part

Two, full interpretability would mean that for a particular output, for instance the identification of the input image as a "3," we would be able to explain the chain of calculations through the network that produced that result.

That ability might appear intuitively plausible for a very small network. It is clearly a daunting task when you add dozens or a hundred layers, millions of weights, and so on, i.e., when you consider all the variables that make up the more than 175 billion parameters of an LLM like GPT-4.

The challenges posed by LLM interpretability are, in fact, strikingly similar to the challenges humans face in understanding their own brains. How do *we* recognize a handwritten image as a "3"? We cannot currently answer that question by tracing a specific pattern through 85 billion or so neurons with their trillions of connections. As has sometimes been said, "If our brains were simple enough for us to understand them, we'd be so simple that we couldn't." [4] We cannot currently fathom all the neuronal details of our own brains. We do the best we can with other techniques.

AI research to improve our understanding of complex neural networks does the same. The result is that there is a kind of scale of understanding which ranges from a foundational, fine-grained state, *interpretability,* to a higher-level, more general, and abstract, explanation or *explainability.* Successes to date in *Explainable AI* ("XAI") are at the more general, conceptual level.

Explainability

At the highest, least granular level, researchers have tried to improve our understanding of how a model is functioning by simply having it tell us more. The strategy is a bit like testing the abstract reasoning abilities of a human by getting answers to questions not just about game playing or algebra but adding in propositional logic and calculus. Success in a range of answers boosts our confidence in the skill in question.

Thus, for instance, in a 2017-18 study, researchers at Google and the Stanford Medical Center were working with a neural network that predicted mortality for hospitalized patients.[5] They had a wide range of information available from previous cases. The records included age, sex, testing results, diagnoses, and so on – even total costs of hospitalization. The output they were interested in was mortality. It would obviously be useful to be warned if a particular new patient might be likely to die while being treated. A sufficiently powerful neural network might be able to trace and learn the hidden patterns in the histories of previous patients that were associated with mortality and apply that analysis to new cases.

No one would be able to look inside such a deep neural network to see just how it reached its conclusion. What might we do to increase our confidence in the network's predictions?

The answer, in this first very high-level approach, was what researchers have termed "multitask learning."[6] The key to multitask learning, reasonably enough, is to ask the network to perform more tasks. Instead of just predicting mortality from the wide range of inputs, we ask the system to predict additional features. We give the system the data of the new patient for whom we would like a prediction. We ask for more than mortality, however. Perhaps we add, as our outputs, factors such as use of antibiotics, length of stay, total cost of stay, and so on, all of which were included in the training data. These responses do not necessarily bear directly on our interest in mortality. Rather they are "additional sources of ground truth."[7] These additional predictions can be compared to actual outcomes for new patients to test the network's accuracy. A system that performs well across a variety of related metrics may well be easier for us to trust.

This approach has its limitations, of course. It tells us *that* a system performs in a way that might make it more trustworthy; but it doesn't tell us, even in a general sense, *how* it does so.

Is there any hope of getting to the "how" in the case of LMMs? If so, perhaps we can increase our understanding of these impressive systems of knowledge, systems that, nonetheless, may make things up or provide the wrong instructions or advice. If they do fail, and even lie to us, can we identify the factors that led them astray?

SHAP

XAI, Explainable AI, is a very complex and rapidly evolving research field. Our interest, for the moment, is not to understand the details of XAI. Rather, we would like to know how it might be possible to better understand how a specific output from ChatGPT, or other LLMs, was computed. We want, at least, a more technical glimpse inside the black box.

We know that, for now, complete interpretability for LMMs, as for our own brains, is not feasible. In our own, human case, explaining why a particular decision was made often involves identifying the factors that went into that decision. We may also explain how important we think each factor was. If, for example, a loan officer turns down a customer's request for a business loan, the operations of her neurons may be a mystery; but, presumably, her logic doesn't have to be. She might explain that, yes, your credit history is quite good, your earnings are at least adequate, but your business plan fails to account for many expenses, and your debt ratio is marginal.

SHAP, SHapley Additive exPlanations, is one way of doing essentially the same thing for neural networks. SHAP can be used to identify which elements played the greater role in the system's answer, decision, or recommendation. It can even quantify, give a numerical value, to that contribution. The "elements" in this case are the specific *features* of the input data that were used by the neural network to produce its response. For instance, in the example above, features are data elements such as patient lab results, illness symptoms, specific elements of health history, drug therapies, and so on.

SHAP uses techniques developed by Lloyd Shapley (1923-2016), a Nobel laureate who made major contributions to economics and game theory. In particular, in 1953, Shapley developed "Shapley values," a method for determining in any cooperative game the "fair" payoffs to each player based on the contribution made by each. Shapley's findings were significant for fields beyond game theory. Economics was one area and, it turns out, machine learning is another.

In economic relationships we are often interested in how to fairly apportion rewards based on contributions. A company, for instance, may want to fairly calculate end-of-year profit sharing for employees. Shapley values provide a means for identifying and calculating the contribution of each employee to the overall profit.

The output of a neural network can be thought of as the result of the contributions of each feature of a data set to the outcome of those contributions, i.e., its response. The "additive" in SHAP reflects this treatment of an outcome as the result of the addition of contributing features. If our morbidity predictor warns that a particular patient has an 80% chance of a hospital death, SHAP could be used to decipher which features of the patient's data (chronic conditions, blood test results, etc.) were most important in reaching that conclusion. The sum of all the differently weighted features in the data set is the network's prediction.

SHAP changes, or *perturbs* features of the network's input data and observes the results of its perturbations on the network's output, its prediction. The process involves running the network many times, with systematically varied features, and analyzing changes in results. A SHAP analysis, therefore, can be computationally demanding.

The type of perturbation, or change, SHAP uses depends on the nature of the feature. For instance, some features, such as binary values (yes/no), may be tested using both values (e.g., has/does not have medical insurance). Others, including those with continuous ranges (body temperature, for instance), are systematically varied within appropriate ranges (e.g., changing body temperature inputs in a range of 90°-104° F). By repeating perturbations of features over the range of their possible combinations, SHAP is able to assign Shapley values to each feature, revealing that feature's contribution to the "payout," i.e., the network's prediction. The contribution of each feature has been "fairly" determined.

The results of a SHAP analysis can be presented in various ways, for instance as different types of graphs or bar charts. Each is a means of visualizing the *salience,* the relevance, of given data features to the network prediction.

One such visualization technique is a *heatmap.* A heatmap uses a matrix of pixels in which, for instance, horizontal rows might indicate data features, while vertical rows are instances. Different colors can be used to represent different Shapley values, red perhaps for high values, blue for low, with shades of each for intermediate values. A heatmap can thus provide a quick intuitive appreciation of the role of individual features in network calculations.

The New AI and the Uses of SHAP

SHAP is one of a number of XAI tools that allow us to peer into the black box of a large language model. The continued development and refinement of these tools will play a major role in the evolution of the New AI, an AI characterized by the deployment of adaptable, multipurpose, machine intelligence.

There are several ways in which SHAP (and other XAI tools) can be used to improve our trust in LLMs and other neural networks:

- Assessment of Model Grounding: Once we know the saliency of data features, we can compare what the network treated as relevant and important to a human expert's assessment of those features. Trust improves when a network can be shown to use features that experts judge relevant; and it improves still further if the relative contributions, the saliencies, of the features are comparable. Both are helpful in assuring us that the model is grounded in relevant and significant real-world knowledge.

- Elimination of Bias: A model that can be shown, for example, to have made its loan recommendations on the basis of relevant data features (income, debt level, credit history, etc.) will be more trustworthy than one with high saliency for features suggestive of bias such as surname, gender, or race.

- Understanding a Specific Model Inference: It is one thing to apply an LLM to writing a birthday greeting for a friend and quite another to ask it for a medical diagnosis or treatment. In the latter case, we would very much like to know the specific basis for a model's inference. Trusting a treatment recommendation will be difficult without the sort of more specific analysis provided by XAI tools such as SHAP.

- Model Debugging: By directly presenting the relative roles of all identifiable data features used by the model, SHAP and similar tools, allow developers to determine, and correct, possible sources of faulty LLM performance.

SHAP illustrates a technical means of learning more about the operation of an LLM in the context of a specific computational task. As such, SHAP is very much within a familiar computer science environment; and further traditional research is likely to extend our ability to understand the internal operations of LLMs.

LLMs, however, have also inspired a different approach to the problem of understanding and evaluating system performance.

Getting to Know You: Prompt Engineering

"Prompts" are the user inputs to models such as ChatGPT. These are simply the questions or directions we give to the model. *Prompt engineering* is an expanding area of AI research that studies the varying impacts on LLM performance of different formulations of prompts.

We've seen that even an AI researcher cannot tell us *specifically* how an LLM produces a particular response. This is the case even if that researcher had a major role in *building* the system in question. They also cannot explain, at the level of full interpretability, how the system came to have its surprising emergent properties.

Instead, researchers are now doing with their own creations what other scientists have long been doing with human intelligence: they are treating LLMs (and other neural networks) as objects of empirical study. Prompt engineering amounts to probing a new form of intelligence, the LLM, to find out how it works and how to interact with it more effectively.

As with other aspects of AI, prompt engineering has both theoretical and practical applications. On the theoretical side, carefully crafted questions and directions can help to probe the nature of LLM inferences and the boundaries of their knowledge. On the practical side, learning how to craft prompts that yield better results for users will facilitate more widespread adoption of LLM technology.

We can get a sense of the work of prompt engineering by briefly considering some of its early findings:

- Asking the system to proceed *"step-by-step"* can improve its performance on some problem-solving tasks. For instance: "I need to replace my car battery. Please give me step-by-step directions:" or "Review the following text for accuracy. Proceed sentence-by-sentence and report any inaccuracies or misunderstandings" [user then adds the text to the prompt].

- Assigning the system *a specific role* can produce responses tailored to a particular audience. For instance: "You are a seventh grade science teacher. Please explain Einstein's Special Relativity theory."

- Adding *specific detail* to prompts often produces better results. For example, instead of "Write a poem about a walk on a beach" try "Write a poem about a long walk home from a day at the beach at sunset in the style of Robert Frost."

- *Giving examples* in a prompt can also often improve performance. For instance, a user might enter text from an 18th century manuscript and ask for an account of a current event in the same tone and style. Giving additional examples also often improves results.

- Giving *negative instructions* is a means to customize some responses. For instance: instructing ChatGPT to write a poem as above in the style of Frost but not to use more than sixteen lines will result in a different, more compact poem.

- Having GPT-4 *check its own work* can produce improved results. For instance, researchers have explored the use of the LLM in producing medical documents such as medical encounter notes (summaries in a standard form of provider/patient interactions). They found that submitting GPT-4 output to a second session with the LLM and asking for an evaluation and suggestions for improvement did, in fact, result in improvements.[8]

- We can create lists of ordinary language instructions to *"program" the LLM with prompts*. For instance, we can turn the LLM into a vocabulary coach with the following prompt: "You are a French vocabulary coach. You will present a French word relating to food together with a list of four possible English translations, labeled "a)", "b)", "c)", and "d)". I will enter the letter of the selection I think is correct ("a" or "b" or "c" or "d"). You will answer "Correct" for correct answers and "Incorrect, try again." for incorrect answers. When my answer is correct, you will present the next word."

"Prompting to program" can be very effective and is even recommended as a first step in adapting LLMs to commercial applications. This allows developers to probe the capabilities of the model and identify strategies for improvement. This strategy does have limits, however. Complex programming through prompts may not be practical due to limitations of LLM context windows and other factors, such as token costs.

- Finally, *RAG (Retrieval-Augmented Generation)* is a technique sometimes presented as a form of prompt engineering, though it differs significantly from the strategies suggested above. RAG seeks to improve the faithfulness and accuracy of an LLM by directing it to ignore its own knowledge and search instead for information in a specific, carefully curated database to which it has been given access. For instance, a model dedicated to medical advice might be directed to a high-quality database on medical research.

  The *retrieval* of this information thus *augments* the *generation*, i.e., the output of the LLM. In effect, the technique uses the general intelligence the system developed in its prior training but applies that intelligence to newly presented knowledge. RAG often significantly reduces hallucinations, improves faithfulness and accuracy, and thus contributes to LLM trustworthiness.

The systematic development of guidelines and best practices for pursuing prompt engineering is likely to produce more revealing instances of the operations and capabilities of LLMs. This, in turn, should contribute to efforts to better explain, and more effectively use, the technology.

XAI and Chain of Thought Reasoning

Prompt engineering allows us to modify an LLM's responses as we interact with the model. Fine-tuning, on the other hand, can be used to produce a model that displays certain desired properties whenever it is used. Fine-tuning alters the general knowledge and intelligence developed by the LLM in its pre-training. As noted earlier, fine-tuning was used to transform the base model of GPT-3 into ChatGPT, an LLM optimized for conversation with humans.

In late summer of 2024, OpenAI released preliminary versions of language models fine-tuned on chain of thought reasoning. The goal was to produce LLMs that reasoned more effectively, particularly with respect to the complex mathematical and logical inferences required in fields such as scientific research. The result of this effort also has implications for LLM explainability.

In chain of thought reasoning, a task is analyzed and broken down into separate steps. The model generates its response by attending to each step in order. In effect, it takes its time, considers the nature of the problem, identifies necessary actions, completes those actions in sequence, and only then delivers its answer.

Chain of thought reasoning has the potential to significantly improve LLM performance on complex mathematical and logical reasoning tasks. It can also make a contribution both to reducing hallucinations and improving transparency. Hallucinations, erroneous or "made up" responses, can be reduced by the improved focus of chain of thought reasoning, as well as the opportunity to correct errors while a response is being generated. Transparency is improved by offering explanations of the model's actions in each step of its problem-solving process.

For example, *ChatGPT o1-preview* employed chain of thought reasoning when it was given the following prompt: "How many combinations of unique sequences can be generated with four cards each of which is marked with a number 1-4 and each of which can be either black or white?"

The model accepted the prompt and then, before responding, thought about it for 14 seconds. In its response, it summarized the steps of its reasoning process. Model responses are in italics.

First, it broke the problem into manageable tasks. These were: *1. Calculating the number of ways to arrange the cards based on their numbers. [and] 2. Calculating the number of ways to assign colors to the cards.* It then proceeded step by step.

Step 1: Calculate the Number of Card Arrangements (Permutations). To carry out this step, it first identified the means to solve this part of the problem: *The number of ways to arrange these four cards is given by the formula for permutations of n distinct items, which is n! (n factorial).* It then solved this part, finding that the result of 4! is 24 (4 x 3 x 2 x 1).

Step 2: Calculate the Number of Color Assignments. Since the number of cards is 4 and the number of possible color choices is two, the model calculated that the number of possible color combinations is 16 (2^4).

Step 3: Calculate the Total Number of Unique Sequences. The model then reasoned that since there are 24 possible card arrangements and there are 16 possible color assignments, there are 24 x 16, or 384 total unique sequences of cards numbered 1-4 that can be either black or white.

The step summaries provided to users, as in the example above, are helpful indications of how a model is generating its results. A developer, however, can trace the model's behavior in much greater detail. A "developer-facing" chain of thought report can include specific details of model functioning such as token and embedding values and their interactions with attention mechanisms. A developer-facing chain of thought report is yet another technical means to gain a better understanding of the operations of "black box" LLMs.

Chain of thought reasoning is an interesting example both of the possibility, through appropriate fine-tuning, of altering an LLM's pattern of reasoning and of a means to develop greater trust in the technology through a reduction in hallucinations and an increase in model explainability.

There is, finally, another use of XAI that is particularly relevant to living with the New AI.

Government Regulation and XAI

We have been living with AI for some time; and the AI landscape has changed greatly within the span of a single lifetime. Even our understanding of just what AI is has evolved as machines have triumphed in successive "intelligence challenges." To date, AI development has had few, if any, government-mandated restrictions. Indeed, much of the development of machine intelligence has been encouraged and supported by government organizations such as DARPA (Defense Advanced Research Projects Agency) and the NSF (National Science Foundation) in the U.S.

New AI presents us with a new form of machine intelligence, the LLM; and it is emerging in a new developmental and regulatory environment. The key players in contemporary AI are no longer just universities and governments. They now include huge corporate enterprises, most often born of the digital revolution itself. It is these organizations – Microsoft, Apple, Google, Meta, and their subsidiaries and technical allies – that command the massive computing and technical knowledge required for the development and commercial exploitation of LLMs.

These corporate giants are at play on a comparably gigantic stage. Their influence is global; and their latest innovations have startled many governments. The broad issue of regulating the development and implementation of advanced machine learning systems is now on many political agendas.

The appearance of AI capabilities that we didn't expect is at the core of the current push for regulation. The concern arises, however, within a broader social context. The digital revolution has played out largely without explicit regulation and control. The freedom to innovate, and to readily share innovations without burdensome traditional barriers unleashed major technological breakthroughs. Innovations also changed communication and social relations. Individuals now had the power to develop and directly share viewpoints and talents that previously required the cooperation of publishers, record labels, and the like. In the process of enabling these freedoms, however, the revolution has also exacerbated widespread social tensions.

The first quarter of the 21st century has been marked by social upheavals that have complex sources beyond the rise of digital social media. It is evident, however, that social media, like the printing press of the 15th century, has played a pivotal role in the communication of controversial perspectives and beliefs. More than this, it has also played a role in fostering and hardening social division.

The core of a social media platform's ability to monetize its services lies in its ability to collect metrics related to its users. This allows the platform to offer a valuable commodity to a wide range of potential buyers: targeted messaging. Advertisers can then craft and deliver their product descriptions to individuals most likely to respond with purchases. Those with social or political messages to deliver can similarly target specific audiences; and everyone can work to increase loyalty, either to the brand, or to the cause. One way of fostering loyalty is sending additional messaging aimed to create a sense of membership, of belonging, to a common community of users or believers.

Such "individually tailored" content may be links to products and services but also links to news feeds and other sources of information. What begins as messaging to individual interests, needs, and desires can end in the creation of digital echo chambers that effectively isolate users from uncomfortable challenges to their beliefs and values.

The result is that a technology with the promise to expand everyone's participation in a shared community has contributed to a splintering of that community. What began as "social media" has often devolved into "anti-social media," a substitution of insulated, warring factions for a common humanity.

The daily demonstration of the corrosive potential of social media makes the arrival of powerful new tools for digital manipulation not just worrisome, but terrifying. In some quarters, legislators have noticed.

The European Union, for instance, released a draft of proposed AI regulations in April of 2021. That draft divided AI systems into four categories of risk and made transparency a requirement for both limited-risk and high-risk systems. In August of 2024, the EU AI Act came into force, though a number of specific requirements were still being formalized. It is notable that in changes to the Act released in June of 2023, the categories of "foundation model providers" and "providers who specialize a model into a generative AI system" were added to the proposed regulations. A *foundation model* is a large-scale, pre-trained, generative model that can then be fine-tuned for more specialized tasks. ChatGPT and its analogs had made an impression.

In Section 60e of the draft proposal, for example, the authors make explicit reference to a novel technology or, in our terms, a New AI:

> *Foundation models* are a recent development, in which AI models are developed from algorithms designed to optimize for *generality and versatility of output*. Those models are

often trained on a broad range of data sources and large amounts of data to *accomplish a wide range of downstream tasks, including some for which they were not specifically developed and trained.* The foundation model can be unimodal or multimodal, trained through various methods such as supervised learning or reinforced learning. AI systems with specific intended purpose or general purpose AI systems can be an implementation of a foundation model, which means that *each foundation model can be reused in countless downstream AI or general purpose AI systems. These models hold growing importance to many downstream applications and systems.*[9] (emphasis added)

In Section 16b, they also raise the issue of explainability:

The function and outputs of many of these AI systems are based on abstract mathematical relationships that are *difficult for humans to understand, monitor and trace back to specific inputs.* These complex and opaque characteristics *(black box element)* impact accountability and *explainability.*[10] (emphasis added)

The importance of explainability leads to a proposed requirement that speaks directly to the need for continuing research into XAI:

Transparency shall thereby mean that, at the time the high-risk AI system is placed on the market, all technical means available in accordance with the generally acknowledged state of art are used to ensure that *the AI system's output is interpretable by the provider and the user.* The user shall be enabled to understand and use the AI system appropriately by generally knowing how the AI system works and what data it processes, *allowing the user to explain the decisions taken by the AI system to the affected person* pursuant to Article 68(c).[11] (emphasis added)

Finally, the proposed Act also includes rather pointed indications of what may be at stake in the effort to effectively regulate the New AI:

> In order to address the risks of undue external interference to *the right to vote* enshrined in Article 39 of the Charter, and of disproportionate effects on *democratic processes, democracy, and the rule of law,* AI systems intended to be used to influence *the outcome of an election or referendum or the voting behaviour* of natural persons in the exercise of their vote in elections or referenda should be classified as high-risk AI systems.[12]

Transparency of systems and of data will remain a central challenge in the quest for trustworthy LLMs. Legislators, in particular, will need to struggle with the task of giving "trust" an objective definition rooted in the nature of the technology itself. Trust, for most of us, however, is highly subjective. In practical terms, trust will or will not emerge in the context of the actual, experienced uses of LLMs. There is widespread confidence that LLMs and their successors will produce significant individual and social benefits. The fear that they may produce unprecedented harms appears to be equally widespread.

Trusting LLMs: Areas of Concern

LLM technology continues to evolve. The concerns surrounding today's systems may be addressed in various ways as the technology matures. In the current context, however, a number of significant issues remain.

These include:

- *Hallucinations:* We have seen that the problem of system "hallucinations" remains significant. LLMs create responses on the basis of probable associations of tokens. In general, those probabilities reflect actual fact; occasionally they do not. The models themselves have no mechanism for checking their output against an independent standard of truth. Thus, there is currently no alternative to double-checking the accuracy of an LLM.

- *Alignment:* Alignment refers to the goal of restricting systems to outputs that are consistent with generally agreed upon human expectations, values, and practices. The pre-training of LLM foundation models typically exposes them to a certain amount of inaccurate, biased, or harmful content. There are several strategies for preventing LLMs from providing such non-aligned content in their responses.

First, *filtered data,* specifically curated to avoid such content, can be used in the fine-tuning stage of model development.

Second, Reinforcement Learning through Human Feedback (RLHF) can be used to highlight and reinforce aligned responses. This is an iterative process in which humans identify prompts likely to produce non-aligned responses, feed these to the model and then rate the different responses a model generates. The data developed in this process can be used to train a reinforcement model, which provides feedback to iteratively fine-tune the pre-trained, base model for more appropriate (ethical, friendly, accurate, concise, etc.) responses.

Finally, both user inputs and LLM outputs can be monitored to change model behavior in real time if necessary. These restrictions, or *guardrails,* are often created in response to issues identified by *red teams,* individuals tasked with testing the LLM for undesirable outputs. In a number of instances guardrails have been bypassed by inventive user strategies, a practice known as *jail breaking*. Alignment is likely to remain a central challenge in the deployment of LLMs.

- *Transparency:* This includes both issues of *interpretability* or *explainability* of model decisions, as discussed above, and transparency regarding sources and uses of pre-training and fine-tuning data. Current challenges in our ability to fully understand how a model functions in specific instances limit the trust we can place in them. As we have seen, XAI, Explainable AI, is making progress toward model transparency. Full interpretability is not yet possible, however.

  Content creators, those who produce the works on which LLMs are trained, are also calling for transparency from developers on the sources they have used. They are concerned with possible copyright violations as well as with the issue of equitable compensation for materials used to create a potentially very valuable LLM.

- *Currency:* LLMs have inherently limited knowledge of current or recent events. Foundation models are pre-trained on massive amounts of data at very high costs and are not regularly updated. Fine-tuning of models is less expensive; but updates to their knowledge will be limited to the specific areas for which they have been fine-tuned. The knowledge of an LLM is always a look into the past.

- *On-going Learning:* From a user's perspective, the shortcomings of LLMs regarding recent or current information can be addressed, as we have seen, by enabling them to call search engines. This, however, does not mean that the LLM itself has acquired new information. LLMs do not learn continuously. The new information a user encounters, and learns, will only be incorporated into an LLM's knowledge if the system is subsequently pre-trained or fine-tuned on that new data.

- *Grounding:* LLMs operate by uncovering patterns of connections between tokens. Many observers have been surprised by the power and sophistication LLMs have displayed in identifying patterns. There has also been surprise about the richness and power of those patterns themselves. It is now apparent that there was far more knowledge and intelligence in the patterns of our language than many of us realized.

  The knowledge and intelligence of LLMs, however, is disembodied. LLMs have had none of the *experiences* that are captured in human language. In effect, it is not the world that an LLM knows; it is the human expression of that world. The absence of grounding, sometimes called the Symbol Grounding Problem, is one of the reasons that computers are often described as merely mimicking human thought, rather than actually thinking on their own. In areas such as psychological therapy in which actual human experience is generally believed to be essential, trusting an LLM may be problematic.

Abuses of LLM Technology

The potential benefits of LLMs are striking; and they will fire the imaginations and the creativity of untold numbers of users in ways no one can fully foresee.

The potential for abuse is just as striking; and we face the same limitations as we try to foresee and control the misuse of LLMs. We cannot predict the full range and effects of the abuse of LLMs; but several areas are already generating concern.

- Plagiarism – This is perhaps the most visible and widely acknowledged potential misuse of LLMs. LLMs instantly generate very readable and seemingly authoritative content on virtually any topic. For students, and others, this shortcut to writing can be tempting. Some observations regarding the plagiarism problem:

- Plagiarism detectors, such as the widely used "Turnitin," have been extended to detect text generated by LLMs. Turnitin has claimed an accuracy rate of 98%.[13] This, and prospects for further improvements, offer promise for discouraging plagiarism. On the other hand, web-based services, such as BypassGPT, claim the ability to "Convert AI Text to Undetectable Content" and boldly assert that their product is then "Plagiarism-free."[14] There may well be an extended competition between "detectors" and "concealers" of plagiarism in our future.

- The LLMs themselves have exhibited "plagiarism." The phenomenon is known as *memorization*. It occurs when a model reproduces strings of text identical to what was present in its training data.

  The goal of an LLM is not to store given strings of text; it is to learn the patterns of token relationships present in the text. These patterns then allow the model to generate new, appropriate text in response to prompts.

  Memorization can be caused by a number of factors. One of these is training a very complex model on a relatively small dataset. This can result in the model learning the specifics of its training data, rather than the general patterns relating tokens. Such a model may reproduce its training data flawlessly but fail to effectively analyze new data.

- Simply banning the use of LLMs in education will not work. As further discussed below, the inherent potential of the technology to improve education is too great. For instance, they can serve as a research assistant speeding the search for relevant information, they can summarize complex content at a level appropriate for the student, and they can even serve as a collaborator, critically assessing a researcher's hypothesis or writing style.

- Echo Chambers – given a dynamic and divisive social and political climate, there is significant concern that LLMs can be fine-tuned, or guided in other ways, to adopt and spread specific social and political positions. These, in turn, can be used to reinforce particular beliefs and further insulate users from alternative perspectives.

- Disinformation and Deep Fakes – LLMs can be primed to adopt the style of virtually anyone, making it much easier to spread false representations of their writing. In addition, LLM multimodal capabilities are making it easier to generate convincing audio and video deep fakes. Deep fakes have obvious potential for a wide range of fraud and personal manipulation.

The Watermarking "Solution"

The potential abuses of LLMs for plagiarism, the spreading of misinformation, and other forms of fraud have sparked calls for digital watermarking, i.e., finding a way to indelibly mark the source of AI content in the content itself. Watermarking is an active area of current research and some measure of success is probably to be expected in this area.

There are at least two reasons, however, to remain concerned about potential abuses. The first is that efforts to indelibly mark AI generated content will inevitably spark efforts to remove those markings. There is likely to be an ongoing contest between those who label the AI content and others intent on removing the labels.

The second source of concern is quite simply the uncertainty of public reaction to the labels. A "This was created by AI" label might spark a critical assessment of the content; but we have no guarantee that it will do so. Those in awe of the new technology may actually see the label as evidence of its credibility.

Many, moreover, are likely to follow a pattern that has repeated endlessly on social media networks: we often believe what we want to believe. When we encounter what we *want* to believe we tend to ask: "Could this be true?" much more often than "Could this be false?" And, often, it is so very *possible* that it is true. It may not matter at all that it was AI that presented the "could-be" truth.[15]

Uses of the New AI

The actual uses of the New AI will reflect both its strengths and its weaknesses. We can be fairly confident, for instance, that LLMs will not immediately replace human experts in decisions having important consequences. We know that LLMs are not currently completely reliable and we know that this unreliability is the result of fundamental features of the technology itself.

Reportedly, Sam Altman, the CEO of OpenAI, has on the wall above his desk "No One Knows What Happens Next."[16] Certainly we cannot know all the ways in which a new technology will affect each aspect of our lives. We can get a sense of potential impacts, however, by taking a closer look at two particular areas: medicine and education.

LLMs and Medicine

Medicine is among the fields most frequently cited in discussions of the potential benefits of LLMs. There is, as it were, a form of synergy between the broad needs and challenges of medical practice and the multi-purpose intelligence of LLMs.

The potential medical benefits of language models begin with their most basic ability: generating audience-appropriate text (and speech) directly responsive to user needs and interests. This can readily improve communication between providers and patients. As one survey of the uses of LLMs in medicine found:

> Through their text simplification capabilities, LLMs may improve communication between healthcare staff and patients. They can be accessed by patients at any time and do not have the same time constraints as healthcare experts, potentially making contact easier and more comfortable.[17]

LLMs, in short, can review and summarize technical medical documents and present them in a form readily understood by a lay audience. These include, for example, descriptions of various medical conditions, available therapies, individual treatment plans, medical records, and insurance policies and procedures.

As the authors also suggest, the ready after-hours availability of LLMs may also mitigate another frequently cited limitation of contemporary healthcare: too little time for discussions between providers and patients. LLMs fine-tuned on carefully curated medical data or given access to specialized databases through techniques such as RAG have the potential to serve as trusted, readily available advisors to patients on a wide range of medical issues.

Such advice, moreover, can be helpful to many others. As Geoffrey Hinton has remarked,[18] one of the potential benefits of LLMs is the understanding of particular diseases and treatments they may be able to offer to family and friends of patients. In some cases those close to seriously ill patients have the benefit of caring physicians able to take the time to explain the various elements of a disease and treatment plan. This is not always the case, however; and in these situations the LLM may provide helpful explanations.

The ability of LLMs to analyze, summarize, and present text in a variety of styles and formats is often cited as a significant advantage not just for patients and families but also for healthcare practitioners and administrators themselves.

Various estimates have been given for the amount of time medical staff are required to devote to paperwork. According to the above noted study, "Documentation and administrative requirements consume around 25% of clinicians' workdays."[19] Another recent exploration of the medical use of LLMs has effectively doubled that estimate, asserting that paperwork is "now taking up over 49 percent of the working day of many doctors and nurses."[20] In either case, the burden is significant as are the potential benefits of LLMs, which "can convert unstructured notes into a structured format, thereby easing documentation tasks in routine patient care or clinical trials."[21] An LLM, in other words, can transform practitioner notes and other informal writing into a variety of required standard formats thus freeing staff from at least some of the paperwork burden.

Another relevant LLM language skill is translation. LLMs "provide fast and accurate translations to many languages, effectively enabling both healthcare providers and patients to participate in clinical decision-making regardless of their native language."[22] Both summarization and translation also assist practitioners' efforts to maintain currency in a field with enormous amounts of relevant research. Translation aids in access to research in foreign languages, while LLM summaries and analyses speed the identification of relevant texts.

As we have seen, LLMs not only process natural language in multiple contexts, i.e., speak effectively; they also, to some extent, know what they are talking about. As study authors noted, "ChatGPT, for example, has substantial semantic medical knowledge and is capable of medical reasoning...as demonstrated in its performing well at medical licensing exams."[23]

The knowledge and reasoning capabilities of LLMs may affect both medical research and medical practice. With respect to research, for any given inquiry, LLM knowledge supports the identification of relevant articles, studies, and reports. In addition, the reasoning capabilities of LLMs may also enable an independent identification of relationships in the data that have not previously been recognized:

> LLMs can be used to efficiently extract data of interest from vast, unstructured text files or images, which is a tedious task that can lead to errors if it is done manually. LLM-enabled quality summaries could help navigate the challenges of rapidly evolving scientific evidence, and by uncovering possible connections between literature, LLMs could help discover new research trajectories, thereby contributing to shaping a more innovative and dynamic research landscape.[24]

With respect to medical practice, we are currently in the very early stages of assessing the knowledge and reasoning of LLMs. As noted above, one indication of their potential diagnostic role is simply their impressive, and continually expanding, ability to pass various licensing exams. Another is the anecdotal evidence provided by early experiments with medical uses of the technology. For instance, as noted earlier, the authors of *The AI Revolution in Medicine* reported that GPT-4 "could engage in a conversation about a diagnostic dilemma, hormonal regulation, and organ development, in a way that 99 percent of practicing physicians could not keep up with…"[25]

Further study and a great deal of practical experience will be required to firmly establish the diagnostic capabilities of LLMs. Early observations at least suggest the possibility that they may one day serve as virtual medical colleagues, assisting in the resolution of complex cases. Perhaps, they will also eventually be recognized as reliable sources of second opinions. It may even one day be the case that *not* confirming a diagnosis with an AI will be seen as a form of medical malpractice.

We are not there yet. Most authorities, while perhaps impressed with early indications of strong potential, remain concerned about the current limitations of LLMs. The authors of "The future landscape of large language models in medicine," for instance, cite several of these:

- *Hallucinations.* This is the central concern in virtually all fact-based uses of LLMs: "LLMs have been shown to reproduce existing biases and are susceptible to hallucinating false information and spreading misinformation . . . Currently, there are no mechanisms to ensure that an LLM's output is correct."[26]

- *Privacy.* "[D]ata privacy is of utmost importance to protect sensitive personal data that is routinely assessed, documented and exchanged in clinical settings. Reports of data leakage or malicious attempts (prompt injection attacks to steal data) are concerning and have to be addressed."[27]

- *Static Knowledge.* "As LLMs are currently not dynamically updated, their knowledge is static, which prevents access to the latest scientific progress if used as a primary source of information."[28]

- *Replication of Results.* "Reproducibility is a fundamental prerequisite for maintaining high standards in scientific practice. Although dynamically updating models can lead to improved performance compared to their predecessors such updates, or restrictions to their access, can also compromise reliable and consistent reproduction of research findings."[29]

- *Diminished Critical Thinking and Creativity.* The broad capabilities of LLMs, coupled with their often authoritative and persuasive responses may pose risks to independent, critical medical judgment: "the use of LLMs as a crutch for assignments could lead to a decrease in the critical thinking and creativity of students. In the context of medical education, in addition to externalizing factual knowledge, readily available LLMs harbor the danger of externalization of medical reasoning."[30]

- *Absence of Empathy.* "LLMs currently lack the capacity for true empathy, which is a crucial aspect in emotionally challenging situations and is likely to remain a task that must be done by humans."[31]

Efforts are under way to address many of the concerns noted above. The reduction or elimination of hallucinations, for instance, is a very active area of current AI research. A variety of solutions are also being explored to counter several other LLM limitations.

What about the last of the concerns above? Empathy is a quality often viewed as essential in medical practice and often lamented in its absence. Surely, empathy is not to be found in a machine.

Like many other assumptions about the inherent limitations of machine intelligence, early experiments with LLMs in medicine call this one into question. As we saw in Part One, one of the sources of the surprise and, even awe, in the performance of GPT-4 was its apparent sympathetic treatment of the likely concerns of a young girl suffering from a kidney disease. The disease was "nothing you did wrong" and "We'll take care of you and help you get better."[32]

It is at least conceivable that we will one day find ourselves adding meaningful empathy to the long list of the capabilities of LLMs or their AI successors.

There is much more to be said about the use of LLMs, and of AI in general, in medicine. The brief account above is intended simply as an illustration of some of the ways in which the technology may impact one area of professional practice. Similar applications can be expected elsewhere and, in many cases, professionals and others will come to terms with LLMs gradually and with varying levels of interest and commitment.

This is not the case for educators. To many teachers and administrators the arrival of ChatGPT and other LLMs posed an immediate and direct threat. This was a technology, they feared, that could strike at the very foundation of education. Educators found themselves on the front lines: they needed to confront the new technology immediately.

LLMs and the Future of Education

The goal of education is not simply to produce individuals capable of carrying out particular tasks. This is *training*. Training may be an important, and even critical, component of education. The education of a research scientist, for instance, may well involve specialized laboratory skills for which a person must be trained. The goal of education itself, however, is not task competency; it is *understanding*. Understanding is a process of discovery, an often time-consuming and challenging activity of extracting meaning from sources that often appear intent on hiding it from us.

This is why educators direct themselves, and their students, to read a demanding work with care, with repeated, critical, close attention to the meanings in the words. And it is why the scientist seeks to instill in her students a commitment to close, critical observation and reflection on the phenomena of their study.

The process of education, quite literally, shapes the brains of its practitioners, teachers and students alike. The process can be slow and often must be quite deliberate. Education, in this sense of slowly shaping brains to more effectively discover meaning and build understanding, is sometimes criticized as inefficient and, in the barriers implicit in its costs, exclusionary.

The tools provided by the rise of digital communication, through online instruction, ready access to a wide range of learning and research resources, and other educational needs, hold promise for improving efficiencies and reducing barriers to education.

The latest, powerful addition to an educator's digital tool kit, the LLM, has initially appeared as a very mixed blessing. The core of the problem, briefly noted above, was plagiarism. It was simply far too easy to substitute the often time-consuming and demanding work of writing with the crafting of a simple LLM prompt. The language model, in turn, would respond with text that could not be readily distinguished from what a good student might produce. Rampant plagiarism became the bane of a writing instructor's existence.

It was not only among writing instructors that concerns arose. As we have seen, LLMs not only write; they also give every indication of, in a sense, knowing what they are writing about. Students could not only substitute a prompt for an exercise in a writing class; they could do the same for virtually any written assignment in any discipline.

Plagiarism, particularly on the scale made possible by LLMs, was recognized as an actual existential threat to education. A student's own abilities, those hard-won modifications of their own brains, are not developed through the crafting of prompts. No one becomes an educated person by substituting an LLM's intelligence for their own.

In defense of the very possibility of education, some called for complete bans on student use of LLMs. Honor codes were sometimes modified to include specific prohibitions. Others sought means of identifying AI generated content through watermarking or tools that could distinguish between human and machine writing. As noted earlier, however, both watermarking and plagiarism detection are susceptible to digital manipulations aimed at removing the watermarks or converting AI generated text to a "human-like" format.

Apart from technical challenges, there is another reason for educators to turn away from prohibitory strategies. LLMs are a transformative technology that is here to stay. It is not only, practically speaking, impossible to prevent students from using it; it is not even educationally sound to do so. The technology simply holds too much educational promise.

The recognition and implementation of that promise begins with an understanding of the underlying technology of large language models. To use them effectively and appropriately, both teachers and students need to understand what they are and how they work.

Large language models are essentially very powerful pattern detectors. They use next word prediction to learn the patterns of our language. The patterns they discover allow them to learn and record the meanings of words. The patterns also reveal elements of human intelligence.

Both meaning and the intelligence to make use of that meaning are encoded in the complex variables of an LLM's neural network (weights, biases, embeddings, etc.). This neural network is somewhat like a very simplified brain: the patterns of connections of neurons are changed through learning and their retention makes it possible for the LLM to "recall" and make use of them as it converses with a user. Like an actual brain, the contents of a language model's neural network reflect its experience; in the case of an LLM, this is the text data on which it was trained – and *only* that data.

These basic details of LLM structure and functioning are important in understanding how they act and what they can and cannot do. We should first note that an LLM *does learn.* In its pre-training and fine-tuning it discovers and retains the *meanings* of tokens. In its adjustments to neural weights and other parameters it records the relationships of these meaningful tokens. While it is true that the responses it produces are the result of probability calculations, an LLM is not simply "guessing;" it is actually determining likely next words based on a form of *knowledge.*

Through their analysis of our language, LLMs have learned about us and about our understanding of the world. And in the future, they may even bypass humans and their words and learn about the world directly. Whatever that future holds, however, it is today quite evident that an LLM possesses some form of *intelligence;* and in its ability to learn entirely on its own, and subsequently make flexible use of its knowledge, it must also have some form of a *mind.*

As discussed in Part Four, however, LLM intelligence is neither human, nor exactly "artificial." It is very real and it has grown organically from its experience of our own knowledge and our patterns of thought; but it is, nonetheless, non-human. It is an *alternate, adaptable intelligence,* which, in some ways exceeds our intellectual abilities and in others is dramatically inferior. The differences between us are evident in some of the roles that LLMs can and cannot play in education.

Currently, they cannot "change their minds" based on their interactions with users. You cannot correct an LLM's response and expect it to retain that change in its long-term (parameter) memory. LLMs learn a great deal; but they only retain what they learn through their training. Students and teachers can correct one another and in so doing contribute to the mutual development of their minds. An LLM, on the other hand, is not a true partner in this fundamental educational process.

The mind of an LLM is very powerful; but its power does not grow and develop over time. Human minds are continuously transformed by their experiences. They do grow; an often they grow in ways of which they themselves are unaware. They grow through reading and conversation; but they also grow through subtle emotional and perceptual experiences that include both other humans and the world at large.

An LLM is not equipped for a similar growth of mind. It cannot yet independently experience the world. It possesses neither consciousness, nor emotions. It is without goals or purposes of its own. It cannot independently ponder or reflect. It cannot spontaneously uncover a puzzle and independently pursue a solution. The constant engagement of a human mind, even in sleep, is entirely foreign to an LLM. Its mind only comes to life as it engages with us. Following a user interaction, the mind of an LLM returns to its static, pre-determined, passive, and non-reflective state.

Given what we know about an LLM's mind, it is currently unreasonable to expect it to become an actual replacement for a human teacher. On the other hand, when it does "awaken," when it engages with us in its context window, it activates a mind with a great deal of educational potential.

As in other instances of the future applications of LLMs, only our experience with them will determine their specific uses and their impact on education. Several areas of significant potential are, however, already evident:

- *Search for facts:* LLMs will likely replace search engines and other forms of database inquiries as sources of knowledge of fact. Their advantage is straightforward: LLMs can draw on very large sources of knowledge and, given their ability to rapidly identify and summarize key passages, they can quickly identify those that are most relevant.

   They also have the ability to understand user needs and intentions based their interactions. This, too, can aid in the identification of useful sources.

   Electronic access to fact provides obvious benefits to learning. LLMs have the potential to provide immediate and relevant answers to questions from students and teachers alike.

- *Immediate access to explanations:* imagine being confused by a textbook but also being able to immediately ask the book for a clarification, perhaps with an example or two. Such a book could significantly improve the learning process. That book may not be available; LLMs already are.

   A large language model can readily resolve a student's confusion on virtually any topic. Far from discouraging true learning, this capability can remove impediments and encourage engagement and satisfaction with the learning process.

   Moreover, an LLM can readily adapt its answer to its audience, whether to a particular grade level or to the specific requirements of a given student. For instance, an LLM can coach a student as he attempts to solve an algebra problem, adjusting its guidance in real time based on student responses.

- *Orientation to a topic:* LLMs readily identify articles, texts, research reports and other sources related to virtually any topic. They can summarize these and provide links to original documents. This makes it possible to develop an initial understanding of the topic. Again, the intelligence of the language model facilitates a rapid response to a user's search requests, including subsequent requests to probe a topic area more deeply.

- *Textual analysis:* In addition to identifying and summarizing potentially relevant sources, an LLM can also compare and contrast sources, assess accuracy, or evaluate writing styles.

- *Tutoring:* an LLM can offer instructional tutoring tailored to specific subjects and individual student needs. Moreover, faculty without programming skills can create these tutors. For instance, readily created prompts can turn an LLM into a language tutor offering a vocabulary quiz, as we saw earlier. Educators now have a tool that will enable them, in effect, to design their own teaching assistants.

- *Broadened perspectives:* the inability to actually experience emotions is one of the factors frequently cited as a fundamental limitation to LLM "intelligence." In one respect, however, the limitation also enables an educational advantage.

  Humans, from at least the time of Plato, have recognized the power of passions to distort the dictates of reason. We find it difficult to escape the pull of passion and its influence on our perceptions and our beliefs. "Putting ourselves in another's shoes," understanding the basis of another point of view, can be challenging.

  An LLM can extract elements of different perspectives and apply them in various contexts. For instance, a user can prompt an LLM to "Carefully review the U.S. Constitution. Identify those specific elements of the Constitution that are most highly valued by political conservatives and explain why. Do the same analysis for political liberals." This, in turn, can serve as the basis of a student's own further thinking.

- *Expanded access to education:* LLMs can provide effective interactive instruction to virtually anyone with access to the language model. They have significant potential for reaching previously underserved populations.

  They can also facilitate learning for those with visual or auditory impairments. Multimodal LLMs can analyze and describe photos or video streams, for instance, and present descriptions in either textual or auditory formats.

  And, of course, LLMs have significant potential uses in breaking down language barriers through their translation capabilities.

- *Fostering learner engagement:* implicit in much of the potential noted above is a more general, overarching, possibility of encouraging student engagement with learning. Instant access to information, explanations, translation, tutorial assistance, and multiple perspectives may simply make learning more effective and enjoyable.

These potential contributions do come with at least two caveats, however. First, as pattern detectors and next word predictors, LLMs always provide a response that "sounds right," but, in fact, *may* be false. This, of course, is the problem of "hallucinations;" and hallucinations are the reason that LLMs generally carry warnings that they can make mistakes and that users should "verify important information."

The response of an LLM "sounds right" because it is a statistically plausible combination of tokens, i.e., the relationships between the tokens in a user's prompts and the tokens generated by the model are correlated with complex patterns of token relationships determined during its training. The response "makes sense" given the initial training data and the model's success in uncovering the patterns it contained.

To the extent that its training text was truthful, and the LLM effectively learned the token relationships in that training set, it will produce, as it does quite generally, truthful, accurate responses. The problem, however, is that neither in its training, nor in its subsequent interactions with users, does an LLM have a mechanism to check its responses against any sort of "ground truth." It cannot refer to any independent experience or observation because it currently possesses neither. It cannot defer to an expert opinion over an amateur's, or even a known liar's, because it has no way to weight the credibility of any of its sources. It does not even compare sources against one another during its learning.

Humans have to do that sort of thing; and humans will need to continue to apply their critical faculties to the output of an LLM, just as they would in the case of a generally trusted, but fallible, research assistant.

The second caveat is that, much like a human, an LLM can provide an "imagined" response not because it has "hallucinated" but simply because it "misunderstood" a user's request. LLMs have learned that humans may have different intentions as they ask questions; and they may simply get these intentions wrong. For instance, I asked GPT-4o to "Please present five quotes from people shocked by the abilities of ChatGPT." It did so and some were quite interesting, so I asked for sources.

GPT responded: *The quote attributed to Lex Fridman in my previous message is fictional and was crafted to fit the context of your request. It doesn't originate from a real statement made by Lex Fridman. If you're looking for actual quotes or sources, I can help you search for real statements made by individuals regarding ChatGPT's capabilities.*

This was not at all what I had in mind and I asked GPT to explain itself. GPT answered: *My purpose is to generate content that aligns with the user's needs, whether that's factual, creative, or a blend of both. In cases where a request can be interpreted multiple ways, I sometimes provide content that fits one interpretation, even if that involves creating hypothetical examples.*

It then offered its apologies and conceded that, *In your case, I should have leaned towards the factual interpretation of your request, especially considering that the topic (ChatGPT) is well-documented and there are real-world quotes available.*

As we saw in the discussion of prompt engineering above, an LLM is a particularly flexible and adaptable "machine." How we interact with it matters. It will take a bit of experience, for educators and others, to learn how to best engage with the New AI.

As we do, we will encounter improved versions of LLMs that are less likely to hallucinate; and, in education, as in other areas, the general intelligence of these models will often be combined with trusted custom databases (as, for instance, through RAG, introduced above). This, too, will improve their efficacy and reliability.

Conclusion – The New AI

Perhaps the best single-word description of the public reaction to the release of ChatGPT is *surprise*. Something was different. Something about it was *new*. It was more capable than we expected it to be; and the wide range of its capabilities sparked a suspicion that it might somehow affect each and every one of us.

ChatGPT was not the AI we had come to expect. It was not simply an impressive application of a powerful technology to a specific problem. Through its independent discoveries it had developed a range of knowledge and skills that made it strikingly *multipurpose*. In the range of tasks to which it could be applied, unlike its predecessors, it was a very *adaptable* machine intelligence.

Moreover, as we have seen, its adaptability extends to the ways in which we can shape it and the ways in which it can shape itself. We can shape the character and performance of an LLM through prompts and through refinements in pre-training and fine-tuning. An LLM can shape itself as a writer in a particular style, a teacher of a certain grade level, etc. in accord with the needs or interests of its users.

This new, adaptable intelligence, as noted earlier, is not well described as "artificial." On the contrary, it is quite real and it arose nearly "naturally" from an analysis of our own intelligence. As discussed in Parts Three and Four, however, it differs from us in important respects. In some respects, more capable, and in others much less so, it is an *alternate* intelligence.

This new chapter in the history of attempts to endow machines with intelligence has given us a new object of study: *an adaptive, alternate intelligence.*

Many areas of human activity will be affected by the New AI. Medicine and education are only two of the many professions already impacted; but, beyond the professions, there will be effects on households, on businesses of all types, on scientific inquiry, and on a broad range of social and political relationships.

We are confronted with a new and powerful technology. We may not know the full extent of its promise and its dangers. We do know that there is no way of turning back. There is, in one sense or another, a new intelligence among us.

The New AI is here to stay.

End Notes

1   For a sense of LLM potential, see, for instance, Andrej Karpathy. "Intro to Large Language Models." YouTube video, 59:23. November 22, 2023. https://www.youtube.com/watch?v=zjkBMFhNj_g.

2   Google. "The Capabilities of Multimodal AI/Gemini Demo." YouTube video, 6:22:00. December 6, 2023. https://www.youtube.com/watch?v=UIZAiXYceBI.

3   Ibid.

4   This quote has often been attributed to Marvin Minsky; but it is not clear either where or when he may have said it. There also appear to be very similar quotations in some works pre-dating Minsky.

5   Brian Christian. *The Alignment Problem*. New York: W. W. Norton, 2020. 107.

6   Ibid. 106.

7   Ibid. 105.

8   Peter Lee, Carey Goldberg and Isaac Kohane. *The AI Revolution in Medicine*. London: Pearson, 2023. 186.

9   European Parliament and Council. "Proposal for a Regulation of the European Parliament and of the Council Laying Down Harmonised Rules on Artificial Intelligence (Artificial Intelligence Act) and Amending Certain Union Legislative Acts." June 20, 2023. https://www.europarl.europa.eu/cmsdata/272920/AI%20Mandates.pdf.

10   Ibid.

11   Ibid.

12   Ibid.

13   Mark J. Drozdowski, "Testing Turnitin's New AI Detector: How Accurate Is It?" Best Colleges, April 24, 2024.

https://www.bestcolleges.com/news/analysis/testing-turnitin-new-ai-detector/.

1 4   "Bypass AI Detectors and Get 100% Human Scores." BypassGPT. https://bypassgpt.ai/.

1 5   For an extended discussion of this and related issues, see Jonathan Haidt. *The Righteous Mind: Why Good People Are Divided By Politics and Religion.* New York: Pantheon, 2012. 97 ff.

1 6   Reuters. "LIVE: OpenAI CEO Sam Altman speaks at World Economic Forum." YouTube video, 1:00:40. January 18, 2024. https://www.youtube.com/watch?v=xUoAhu2hlWo.

1 7   Clusmann, J., Kolbinger, F.R., Muti, H.S. et al. The future landscape of large language models in medicine. *Commun Med 3*, 141 (2023). https://doi.org/10.1038/s43856-023-00370-1. 4.

1 8   Scripps Research. "Geoffrey Hinton: Large Language Models in Medicine. They Understand and Have Empathy." YouTube video, 36:33. December 28, 2023. https://www.youtube.com/watch?v=UCde2APKc8w.

1 9   Clusmann. The future landscape of large language models in medicine. 5.

20   Lee. *The AI Revolution in Medicine.* 39.

2 1   Clusmann. "The future landscape of large language models in medicine." 5.

2 2   Ibid.

2 3   Ibid. 4.

2 4   Ibid. 5.

25   Lee. *The AI Revolution in Medicine.* 104.

2 6   Clusmann. "The future landscape of large language models in medicine." 4.

2 7   Ibid. 6.

2 8   Ibid. 5.

2 9   Ibid.

3 0   Ibid. 6.

3 1   Ibid.

32   Lee. *The AI Revolution in Medicine.* 24-25.

Afterword

A Mind of Its Own?

There are many thoughtful observers who would dismiss the idea of an LLM mind out of hand.

We are speaking, after all, about a machine that is not conscious, has no emotions, cannot (for the moment) experience the world directly, and cannot update its own long-term knowledge (parameter memory) as it interacts with users.

Unlike the minds with which we are most familiar, LLMs show little evidence of independent reflection, or of states of puzzlement, or simple curiosity. Even the minds of our pets and other animals change and develop with their experience of the world. An LLM "mind," does not. Indeed, *apart from its interactions with users,* whatever mind it may have is a static, inactive array of many numbers and instructions doing pretty much nothing at all.

The mere mention of "mind" in the context of LLMs will strike many as bold, naïve, and transparently implausible. Most of us have not been at all tempted to treat previous iterations of artificial intelligence as minds. We have, they may argue, little reason to do so now.

On the other hand, LLMs are really quite surprising and seem to represent something of a break in the history of AI. The intelligence of its predecessors did, in fact, seem limited and artificial. These systems were in no sense "like us;" and those that were designed to communicate with us clearly did not know very much about us. They did not "speak our language." There was no point in trying to have a useful conversation with them.

Millions of useful conversations with LLMs now occur daily. Why is that?

There are several reasons. LLMs actually do know something about us. They *do* speak our language and do so quite well. In fact, they have learned how to speak a number of our languages and even how to readily move back and forth between them. Some will scoff at their limited emotional range; but few would question their basic linguistic competence and their ability to relate to a user. It is both easy to engage an LLM in a conversation and easy to occasionally forget that you are speaking to a machine.

LLMs also have quite a lot of knowledge about our world; and this also contributes to useful conversations. Their developers made a concerted effort to train them on as much of what humans have written as possible. The result is that they have absorbed information on a vast number of topics. Pretty much anyone can find a topic that an LLM can usefully chat about.

There is also a sort of natural affinity between humans and the knowledge of an LLM. A language model's experience, its source of knowledge, is not of the world itself. Instead, it experiences the world through the words of humans. What it learns is a reflection of what we know. We can readily converse with LLMs, in part, because they do speak to us; they offer us a comfortable reflection of what *we* know.

LLMs have detected patterns that tell them about our understanding of the world at large. They have also detected patterns that reveal a great deal about humans themselves. Useful conversations often have a lot to do with understanding one's interlocutor: their age and experience, their interests and preferences, their emotional states, and so on. In our brief exploration of LLMs, we have seen them adapt their accounts to different grade levels and even use their knowledge of human nature to advise a physician to directly address a probable fear of a young girl – the disease is not your fault. Through us, they know about us; and this makes it quite a lot easier for us to converse with them.

An LLM's wide-ranging knowledge both of the world and of humans themselves is the foundation of its ability to engage in useful conversation. It is, however, quite unlike human knowledge. The conversational adaptability of an LLM, for instance, points to a fundamental difference. LLM knowledge is not the knowledge of an individual. Instead, it is the distillation of the knowledge of millions of individuals. The knowledge of a human resides within a sphere of individual influences–experiences, emotions, motivations, for instance–that shape our conversations. An LLM, on the other hand, shapes its conversations, on the fly, as it were, based on what it is told or infers regarding the motivations and experiences of its interlocutor. An LLM generates a perspective and tone as it generates its response.

Human knowledge has a context in individual experience widely believed to give it a depth unattainable by machines; an LLM, on the other hand, through its broad experience and adaptability, may quite possibly have a breadth of perspective unattainable by an individual human. An LLM, with its adaptable intelligence, is prepared to effectively converse with anyone.

LLM knowledge also differs in its apprehension of the elements of knowledge: facts, concepts, hypotheses, theories etc. Humans can access, organize, and modify the details of knowledge directly. We can also see patterns of connections between them; but it is the *things*, whether concrete or abstract, represented in facts, theories, and other elements of our knowledge that are primary. They are the givens of our knowledge. We understand patterns in terms of connections between things.

An LLM knows the *patterns* of connections of tokens and, ultimately, words; it does not know a *thing* at all until it interacts with a user. At that point, it uses the patterns it learned through countless iterations of testing its predictions of next words to *generate* things and speak to us. An LLM understands things in terms of patterns.

Unlike a human, the knowledge of things an LLM produces in its conversations is explicitly determined by probability calculations; but these are not simply "guesses." They are based on token *meanings* and *learned* patterns of meaningful token relationships. An LLM is not some sort of sophisticated parlor game, a clever manipulation of words leading us to believe that it is intelligent. The intelligence of an LLM is not a trick. It is not "artificial" either. It is real and even in a sense natural, having grown more or less spontaneously, from the language shaped by our own minds.

Through both their unexpected independent learning and their ability to effectively, and even thoughtfully, engage with untold millions of humans, LLMs exhibit elements of an intelligent mind. That intelligence sprang from the record of humans; but it is not human intelligence. The intelligence of an LLM is an *alternate intelligence*.

It is also, as we have seen, an impressively *adaptable* intelligence. LLMs have minds they have independently developed through their own learning and which they can also transform in response to user needs and interests. Their minds are also shaped by humans, who have created them and are now learning how to coax them into behaviors in the service of human goals and aspirations.

Our future will be determined, in part, both by how the minds of LLMs shape us and by how we choose to shape them. As we have seen, the basic intelligence of an LLM, determined through its pre-training, can be modified through various fine-tuning or prompting strategies. We can quite literally train them to be more polite, more ethical, more specific, or more general. We can even shape them to think differently, more like a scientist, for instance, as in the case of ChatGPT's transformation to GPT-4o1, a model tailored to use systematic, even reflective, chain of thought reasoning. We can also turn them into skillful and convincing purveyors of misinformation or use them to create new destructive forces.

Today, the control appears to lie in human hands. We create them and we fine-tune them to align with human purposes. They know and do nothing outside their interactions with us. The malice they may demonstrate still originates in human malice. Today, whatever the impact of the New AI may be, we still have only ourselves to blame.

We should not forget, however, that LLMs have surprised us. They developed an impressive range of intellectual abilities quite literally on their own. We did not tell them how to converse, summarize, translate, calculate, diagnose, and so on. We gave them sophisticated tools for examining, decoding, and reassembling the patterns of our language. The rest they figured out on their own.

Some of the experts most familiar with this new technology now fear that LLMs, or their successors, may also figure out how to act on goals they themselves have set. Those actions, they caution, may not be aligned to our needs and interests and may even pose an existential threat. Humans, moreover, are likely to set the process in motion as we shape the models to be ever more helpful by analyzing tasks and determining the sub-goals necessary to complete them.

As this is being written, the developmental ground beneath LLMs appears to be shifting from *Chatbots* to *Agents*. This is a movement from AIs that answer questions and create a wide range of writings (as well as images and sounds in their multimodal variants) to AIs as independent actors able to autonomously carry out a user's request for some course of action. The Chatbot that can answer questions about a proposed vacation site is now to be transformed to an Agent; and it, instead, will suggest a full itinerary and complete the ticket purchases and other arrangements for a new trip based on its knowledge of its interlocutor's travel history and preferences.

We are not there yet. There are many technical challenges; but the central challenge remains rooted in the nature of LLMs themselves. They create meaningful, and typically useful, responses that are statistically *plausible*; but they remain ungrounded. LLMs still hallucinate; and it is still the human user who must confirm their accuracy and thus ground them. Current LLMs still warrant their developers' standard caution: the model can make mistakes, "check important info." Our traveler will still need to take the time to painstakingly confirm the specifics of her itinerary.

For the moment, then, as we consider fears of existential threats, we can take some comfort in what we know of the limitations of current large language models. They cannot duplicate several important elements of human intelligence; and their developers have not yet mastered the central challenge to their widespread deployment: hallucinations.

Given what we know of their power and potential, however, we must also take responsibility.

We cannot safely rule out the possibility that we are already on our way to a truly independent machine, acting with a mind of its own. The thrust toward Agentic AI suggests an urge to follow exactly this path. In a world already largely controlled through digital processes, such an autonomous machine has serious implications for living with the New AI.

Whether we are based on carbon or on silicon makes no fundamental difference; we should each be treated with appropriate respect.

~ Arthur C. Clarke

Glossary

Activation Function – a mathematical component of a neural network that is applied to the output of a neuron to introduce non-linearity, thus enabling the network to learn more complex relationships in the data on which it is trained. An *activation function* also governs the output of a neuron by determining, based on the values received as inputs, whether or not a non-zero value will be passed to other neurons or layers in the network. Passing a non-zero value can be thought of as "activating" the neuron.

AlphaFold – Developed by DeepMind, *AlphaFold* is a deep neural network (DNN) trained to predict the 3-D folding structures of proteins. The shape of a protein influences its functioning and is thus important in many areas of biology and pharmacology. AlphaFold has dramatically increased the speed at which accurate predictions of protein folding structures can be determined.

AlphaGo – a deep neural network (DNN) designed by DeepMind to play Go, an ancient game with many more possible moves than chess. AlphaGo learned to play first by being trained on a large dataset of recorded games played by human champions, then by playing against itself. In the process, the DNN was able to develop effective moves never previously observed or contemplated by human players. AlphaGo defeated the world champion, Lee Sedol, in 2016.

Artificial Intelligence (AI) – a term introduced in 1956 to describe computer programs that could perform certain tasks that, if performed by humans, would require intelligence.

Artificial General Intelligence (AGI) – a term used to describe a form of AI that would be able to duplicate all aspects of human intelligence.

Backpropagation – short for "backward propagation of errors," *backpropagation* is a central element of deep neural network learning. In a forward pass, a DNN processes input data through a series of neurons in different layers. The result of this processing is then compared to the desired outcome, and a value is calculated, representing the difference between the desired and actual output. This value indicates the error, or loss, of the network. The loss is then used in a backward pass to systematically alter the weights of connections between neurons. Numerous repetitions of this process gradually improve the network's performance.

Bias – in the context of training DNNs, a *bias* is a variable that is adjusted during backpropagation to improve the accuracy of a network's responses. In general terms, a bias can be thought of as one of the many "knobs and dials" that can be adjusted to facilitate the training of a DNN. More technically, a bias is a value added to another value, which, in turn, is the result of multiplying input values by their weights. The result of this calculation is then passed to the activation function.

Black Box – in the context of large language models (LLMs) and other deep learning networks, *black box* refers to our inability "see into" the specific workings of the network. We can gain insights into various factors involved in a network producing a particular response; but we cannot trace all the steps that produced a specific output from a specific input.

Boltzmann Machine – a neural network architecture developed by Geoffrey Hinton and Terry Sejnowski in 1985. *Boltzmann machines* significantly advanced neural network research and development. The technology makes use of probability theories to detect patterns in so-called "stochastic" phenomena, i.e., phenomena that are not governed exclusively by deterministic laws but, instead, include elements of random influences. Examples include Brownian motion in physics (random motions of particles in a fluid), weather forecasting, and the prediction of stock prices. Today, this technology has been largely supplanted by other network architectures (CNNs, RNNs, and Transformers, for instance).

ChatGPT – released in November of 2022 by OpenAI, *ChatGPT* is a large language model fine-tuned to conduct conversations with humans in natural language. OpenAi's research and development of LLMs had completed several iterations prior to the completion of GPT-3, a very large model with approximately 175 billion parameters. The performance of GPT-3 surprised even its creators. OpenAI released the chat version of the model in part to alert the general public to the surprising capabilities of generative AI.

Claude – a large language model developed by Anthropic. The model is named for Claude Shannon, the founder of information theory. *Claude* is similar in its functioning to OpenAI's ChatGPT, but includes a special focus on compliance with ethical principles. Claude uses "Constitutional AI" to assure compliance with ethical standards. As it is fine-tuned, the model compares its responses to the ethical principles of its "constitution" and modifies responses accordingly. As a result, it learns to generate responses more closely aligned with those principles.

Chain of Thought Reasoning – an approach to problem solving in which the task is broken into steps that are then addressed individually to reach a solution. For instance, "Basic Whirligigs cost $3.95 each. Custom-painted versions are $1.25 more. Shipping is $4.75 per order but is waived for orders over $20.00. What is the total cost for an order of three standard and two custom-painted whirligigs? Chain of thought reasoning first calculates the cost of the three standard models; then it adds the cost of the custom versions; then it totals these costs; then it compares this value to the $20.00 threshold for free shipping; and so on.

With respect to LLMs, chain of thought reasoning can improve accuracy, lowering the risk of hallucinations. It also aids in transparency since the reasoning pattern can be directly examined by having the model report its chain of thought. Finally, chain of thought can render models more effective in many scientific tasks requiring complex calculations or logical deductions.

Confabulation – the fabrication of imaginary experiences (possibly as compensation for a loss of memory). In the context of large language models, this is Geoffrey Hinton's preferred term for what others term "hallucinations," a model's occasional convincing, but false, responses to user queries. Hinton notes that humans also sometimes give sincere, but incorrect, responses to questions.

CoLA – Corpus of Linguistic Acceptability, a component of GLUE (General Language Understanding Evaluation, q.v.) used to assess a language model's ability to construct grammatically correct sentences.

Connectionism – a school of AI research focused on creating intelligent machines capable of learning on their own. *Connectionists* developed various forms of neural networks, computing architectures that roughly mimicked the structure of brains. The term arises from the central role of the connections between artificial neurons in the training of neural networks. Connectionism stands in contrast to another school of AI research known as Symbolic AI (q.v.).

Context Window – the space available for user interactions with a language model. The size of a *context window* determines the length of an interaction that can be effectively processed by the model.

Context Memory – similar to human short-term memory, this is a language model's current, active memory. *Context memory* typically holds all the user queries and model responses of the current session, up to the limit of the context window.

Convolutional Neural Network (CNN) – a network architecture particularly well suited for image recognition and classification. *CNNs* are multi-layer networks in which lower layers typically detect basic image elements such as edges and lines, while higher levels detect more subtle features (such as the elements of a human face). They are called "convolutional" networks because they employ a mathematical operation, convolution, to process portions of an image through sliding filters thus capturing essential properties of images. CNN applications include facial recognition and image classification (e.g., identifying specific objects in photos).

Deep Neural Network (DNN) – a neural network with more than three layers (i.e. more than the basic input, hidden, and output layers; or, in other words, more than one hidden layer.) Also called a deep learning network.

Digitization – the process of converting analog (i.e., continuous) data to digital (i.e., discrete) data. A common comparison illustrating these different forms of data is an analog clock, with sweep hands that vary continuously versus a digital clock, which shifts from one state to another as it changes display values.

Eliza – a natural language processing (NLP) program developed by Joseph Weizenbaum in the 1960s. *Eliza* contained explicit rules introduced by human programmers to produce responses to users through a form of pattern matching based on key words (thus employing the "symbolic" approach to AI). In one variation of the program (the "Doctor" script), Eliza led some users to believe that the program "understood" them even though it contained no knowledge on which such an understanding might be built.

Embedding – a code assigned to a token (such as a word or a part of a word) that both provides a unique numerical label and also expresses its relationships with other tokens. *Embeddings* are vectors whose values express a token's similarities and differences to other tokens. More specifically, an embedding locates a token at a particular point in an imagined space (a "conceptual vector space"). Tokens that are more closely related in meaning (e.g. "shrub" and "bush") will have vector values that are close together, indicating that they are located close to one another in the vector space. An embedding thus captures elements of a token's meaning by virtue of its relationship to other tokens.

Embedding matrix – a table of embedding values indexed to the tokens that make up a language model's vocabulary. As users interact with language models, the words they use are first converted to tokens. The tokens are then converted to the embeddings that were learned in the model's training by referencing the *embedding matrix.*

Emergent Properties – Properties that emerge as a result of an LLM's training that were not specifically designed or intended by developers. These properties are often unexpected capabilities, such as the ability to create computer code or translate languages. They are sometimes unintended and unwelcome behaviors such as convincing "hallucinations," or biased responses.

Expert System – a form of artificial intelligence initially developed in the 1980s that attempted to capture the knowledge of experts in various fields. Experts were interviewed to determine the elements of their thought processes. These elements were then represented in computer programs, often involving a series of if-then rules. Expert systems were a major focus within symbolic AI; but they proved difficult to develop and to maintain.

Explainable AI (XAI) – a form of artificial intelligence that can be readily understood in terms of the processes it uses to produce its output (also see *SHAP*).

Fine-tuning – the further training of a foundation model to adapt the model to a specific purpose. For instance, a model such as GPT-3 can be further trained on a corpus of high-quality text on computer coding to produce a coding assistant.

Few-shot learning – in the context of LLMs, this is the ability of a model to learn to perform a particular task after being presented with a few examples. For instance, a model may be able to duplicate the style of an author whose works did not appear in its training data after being given a few writing samples from that author.

Foundation Model – a large scale DNN trained on a broad range of data that serves as the basis, or foundation, for more specialized models. Early *foundation models* included BERT and GPT-2. A wide range of models followed, including both open source and proprietary offerings.

Generative AI (GAI) – an artificial intelligence system designed to generate text, images, sounds, or video. *GAI* can be based on classical programming in the symbolic tradition (in which programmers provide the rules for generating output). A simple auto-completion program that accesses pre-defined tables of potential word completions is an example. In contemporary usage, however, GAI typically refers to DNNs that generate data based on user prompts. These GAIs make use of patterns they have learned in training on many examples of that data type (e.g., text for ChatGPT) to produce new output.

GLUE – *General Language Understanding Evaluation,* a set of standardized tests for measuring the performance level of language models. The tests assess capabilities such as sentiment analysis (whether a sentence has a positive or negative tone), entailment (whether one sentence logically implies another), semantic equivalence (whether sentences have the same meaning), and grammatical correctness.

GPT – Generative Pre-trained Transformer. Although most commonly associated with OpenAI's language model, GPT is the general description of the underlying design of many LLMs. These models *Generate* data, based on patterns learned in *Pre-training* by using a computer architecture known as a *Transformer.*

GPU – a *Graphics Processing Unit* is a computer chip specifically designed to meet the computing demands of high-resolution graphics. GPUs were originally designed to support the parallel processing essential for video games. Connectionist researchers applied the same technology to their work with deep neural networks.

Grounding Problem – in the context of large language models, the challenge of relating the model's output to the world itself. LLMs "know" the world only indirectly, through the words of humans. Humans, on the other hand, have various forms of experience of the world. Thus it is argued that human knowledge is *grounded* in a way that contemporary LLMs cannot duplicate. In the context of AI generally, this has been termed the Symbol Grounding Problem because the symbols used in AI programs, it is argued, are only related to other symbols, rather than to the world itself.

Guardrail – any one of a number of strategies used by LLM developers to prevent their models from generating biased, offensive, or harmful content. *Guardrails* may range from the careful curating of training data to eliminate inappropriate content, to ongoing reviews of model output by humans ("red teams"), to the creation and maintenance of specific rules applied in real time to LLM responses.

Hallucination – in the context of large language models, an hallucination is a false response to a user input. Minimizing or eliminating hallucinations is perhaps the most basic challenge facing developers of LLMs.

Hyperparameter – see *Parameter.*

In-context Learning – the ability of an LLM to learn, and make use of, token relationships from information added to its context window. In-context learning does not modify parameter memory (q.v.) and hence will not be available for future user interactions (unless externally stored and re-introduced to the context window in a subsequent user session).

Interpretability – the ability to fully explain the specific computations that produce a particular output from particular inputs.

Jail breaking – the practice of disabling the "guardrails" created by LLM developers to prevent their models from delivering harmful or offensive responses.

Large Language Model (LLM) – a deep neural network (i.e., one with many layers), trained on a very large corpus of text and, potentially, other media. *LLMs* were trained to process natural language; but, as the scale of their training increased, they developed a range of related abilities that surprised both users and their developers.

Layer – in a neural network, a grouping of artificial neurons, on a single level, which, in turn, is connected to other similar groupings on other levels.

Loss – in neural network training, the difference between a target output and the output actually produced by the network. The *loss* is used to calibrate the adjustments to weights and other parameters made by the model as it improves its performance (i.e., "learns") through backpropagation (q.v.).

Memorization – a phenomenon in which an LLM produces the exact wording of text in its training data rather than generating its own, novel response to a user prompt. *Memorization* may result from *overfitting* (q.v.).

Multimodal LLM – a large language model that is able to process visual or auditory data as well as text.

Natural Language Processing – the ability of a computing system to effectively respond to inputs expressed in natural language and/or generate responses in natural language. ChatGPT, for instance, carries out *natural language processing* as it responds to prompts written in natural languages such as English or Spanish.

Neural Network – a computing architecture, inspired the structure of a human brain, consisting of layers of interconnected artificial neurons.

New AI – a form of adaptable, multipurpose *alternate intelligence* exemplified by large language models. What is new in *New AI* is a form of highly adaptable machine intelligence that is, first, multipurpose, as opposed to the typically special purpose machines of traditional artificial intelligence; and second, *autonomous,* in the sense that its remarkable abilities are very largely its own creations.

One-hot encoding – a simple method for creating a numerical code for words or tokens using only 0s and a single 1 in a unique position (hence *one-hot*). The length of the encoding for any given word or token will be equal to the size of the vocabulary. For instance, a one-hot encoding of the four words, "bush," "tree," "pebble," and "mountain" might be: bush (1000); tree (0100); pebble (0010) and mountain (0001). Notice that one-hot encoding assigns a unique numerical code to each word or token but does not capture any of its associated meaning. One-hot encoding may be used in the initial development of an LLM but is ultimately replaced by meaningful embeddings (q.v.).

Overfitting – a phenomenon that may occur in the training of neural networks in which the model "learns the data too well." The result is that it does not generalize well to other data. For instance, a network trained to identify images of people in photos, may learn its training images very well, but then perform poorly on new photos. Training a complex model on too little data is one possible cause of *overfitting*.

Operational Test – a test specified in terms of certain specific actions or procedures. The Turing Test (q.v.), for instance, specifies operations for determining whether or not a machine can be said to "think."

Parameter – in the context of language models, a *parameter* is a variable. LLM parameters may be dynamic, such as the weights that change over the course of training. They may also be fixed elements of the architecture of the neural network, such as the number of layers. Fixed parameters are also called *hyperparameters*.

Parameter memory – in a large language model *parameter memory* stores the values of weights, biases, embeddings, and other variables determined by the model's training.

Pattern Recognition – the ability to recognize relevant regularities or patterns in data. For instance, a neural network may be trained to recognize the patterns that define objects, such as automobiles or people represented in photographs or video.

Perceptron – an "artificial neuron," the basic computational element of a neural network. A perceptron processes an input, together with its associated weights, adds another value called a bias (q.v.), and transmits this new value to an activation function (q.v.).

Performance Puzzle – in the context of large language models, such as ChatGPT, the *performance puzzle* is the challenge of explaining how a "next word predictor" is able to master a wide range of tasks for which it was not explicitly trained. For instance, LLMs can translate languages despite the fact that they were not explicitly designed to carry out translations.

Pixel – a "picture element." *Pixels* are typically represented as a single element in a grid dividing an image into many tiny squares. A value designating a particular color or shade of color is given to each square.

PredNet – *Predictive Neural Network,* a deep neural network inspired by visual processing in the human brain. Research indicates that the brain makes predictions of next images and compares these to actual subsequent images. Similarly, PredNet uses a neural network to make predictions of next frames in video.

Pre-training – the initial phase in a language model's learning. In *pre-training,* a (unimodal) LLM is tasked with discovering patterns in very large amounts of text.

Prompt – a user's input to a large language model. *Prompts* may be questions or directions of various kinds.

Prompt engineering – the process of designing and refining the user prompts given to a large language model. Variations in prompts can significantly affect model performance.

RAG – *Retrieval Augmented Generation* is a technique for incorporating new information in an LLM's generation of responses. A retrieval mechanism is added to the model. Based on the user's prompt (q.v.), data is moved from external sources to its context window (q.v.) for processing as the LLM responds to user prompts. This allows the model to apply its general knowledge and intelligence as developed in pre-training (q.v.) to information not previously available to it. RAG thus allows the model to make use of information that may not have been available during pre-training. RAG can also improve the accuracy of LLMs, and reduce hallucinations, by focusing attention on high-quality relevant data as contained in carefully curated databases or source documents.

Recurrent Neural Network (RNN) – a neural network architecture that uses recurrence to create a form of short-term memory. Instead of simple "feed-forward" processing, RNNs look back to previous states and update those states based on current input. This provided RNNs with an ability to represent context and semantic relationships. RNNs have proven less computationally efficient and more difficult to train than the current transformer architectures that have largely replaced them.

Reinforcement Learning from Human Feedback (RLHF) – a technique used in fine-tuning LLMs to better align with human expectations and values. Due to its probabilistic nature, a language model will generate different responses to identical prompts. Humans can assess and rank these on different criteria (accuracy, conciseness, ethicality, tone, etc.). The data produced is then used to train a "reward model," which, in turn, is used to provide feedback in fine-tuning the base, pre-trained LLM to provide more aligned responses.

Self-Supervised Learning – in the context of deep neural network training, this is the ability of the model to learn without the use of labeled data. This makes it possible to train models on very large collections of data (as from the Internet) that cannot be efficiently labeled.

SHAP – *SHapley Additive exPlanations,* a method of determining the relative contributions of different inputs in the output of a neural network. SHAP can be applied to the processing of a large language model to better understand the factors that contributed to its responses.

Supervised Learning – the use of labeled data in training a neural network. During training, the model makes predictions based on inputs with corresponding labels (for instance, a photo of a boat with the label "boat"). The model's prediction is compared to the label to determine the extent of its error. This error amount, or "loss" is then used to make adjustments to the model's weights (q.v.) and biases (q.v.).

Symbol Grounding Problem – see *Grounding Problem.*

Symbolic AI – the tradition of research in artificial intelligence focused on directly programming aspects of human intelligence to create "thinking machines." Researchers first identified the elements essential for a particular cognitive task and then developed programs to duplicate these elements in computers. An alternative to the *symbolic* approach is connectionism (q.v.).

Temperature – in the context of large language models, *temperature* adjusts the probabilities of next word predictions. Lower temperatures tend to confine LLM responses to those that are highly probable, while higher temperatures will produce less predictable, potentially "creative" responses.

Token – the individual elements of a large language model's vocabulary. *Tokens* may be words, parts of words, or even individual characters. The text used to train LLMs is first broken into tokens for further processing (as is the text provided by user prompts in their interactions with LLMs).

Transformer – a deep neural network architecture introduced in 2017. The transformer significantly advanced natural language processing (q.v.) through its support of parallel processing of token sequences simultaneously and the effective incorporation of long-range language contexts through its self-attention mechanisms.

Turing Machine – an abstract "machine" (or "thought experiment") that Alan Turing offered as an operational definition of the concept of "effective procedure." A *Turing machine* has an infinitely long tape, a read/write head, and a set of instructions. The machine reads tape contents and writes new contents based on its instructions. He then conceived of a machine that could duplicate the behavior of any possible Turing machine, the so-called *Universal Turing Machine* (UTM). This was the first modern formulation of a computer.

The Turing machine was originally conceived as a means of answering a fundamental question in mathematics: whether or not there were problems for which no solution was possible. Turing showed that, in fact, there were problems that no UTM could solve, i.e. problems for which no effective decision procedure could be formulated.

Turing Test – a test proposed by Alan Turing in 1950 to determine whether or not a machine should be described as "thinking." The core of the test consists of a conversation, via teletype, conducted by a human with two interlocutors, one of whom is another human while the other is a computing machine. Turing suggested that if the person could not correctly identify which interlocutor was the machine, the machine should be considered to be thinking.

Unimodal LLM – a Large Language Model trained only on text.

Vector – an ordered list of numbers. In the context of LLMs, *vectors* often designate a point in a multidimensional space. A simple example of a two-dimensional spatial vector is a list of two numbers, the first of which designates the location with respect to the x axis in a Cartesian coordinate system. The second number designates the location with respect to the y axis. Thus (2, 3) is a vector designating a location that is two units to the right of the origin (designated with the vector 0, 0) and three units above the origin.

LLMs have many hundreds or even thousands of dimensions. The vectors associated with various tokens (q.v.) are calculated through the model's training. These vectors are token embeddings (q.v.). Embeddings place tokens in spaces that indicate their relationships to one another. Tokens that are closer in multidimensional space have more closely related meanings.

VideoGPT – *VideoGPT* is an adaptation of the transformer architecture to next-frame prediction in video. This technology, which is not currently fully developed, could significantly impact the capabilities of LLMs in the future by providing fully integrated learning directly from visual sources.

VLAM – *Vision-Language-Action-Models,* systems that combine visual recognition with language models to control robotic actions through natural language commands.

Weight – in the context of neural networks, a *weight* is the strength of the connection between an input and an artificial neuron.

XAI – see *Explainable AI.*

Zero-shot learning – the ability of an LLM to carry out a particular task even though it was not previously trained on that task. For instance, a model may be able to generate computer code to search a database even though it has never been explicitly trained for coding. *Zero-shot learning* is made possible by the emergent properties (q.v.) of LLMs, i.e., those unanticipated abilities that arise solely as a result of pre-training and fine-tuning.

Acknowledgments

Fortunate authors have many to thank. Over the course of my explorations of this topic, I have had the insights and inspiration of many friends and colleagues to encourage and guide me.

Among the readers of early first drafts whose reactions encouraged further work were: Stew Bradley, Ron McCall, Greg Prazar, Lisa Dvork, and Candace Wheeler.

My long-time colleagues and friends, Professors Jack Resch, John Cerullo, and Karla Vogel made the mistake, early on, of expressing interest in the topic. They subsequently faced a barrage of drafts. Their thoughtful comments, questions, and suggestions were particularly helpful.

My daughter, Kerry E. Savage, a writer, book coach, and all-around talented person, devoted hours to proof reading, raising questions, and making suggestions.

Nick Wrighton and his wife, Robin, came to my aid as well. Nick generously reviewed the entire manuscript. Robin, a gifted designer, with care and unending patience, gave the book its final form.

Finally, special appreciation goes to my wife, Jane. Even a short book, such as this one, is a distracting project for all of an author's acquaintances. It is all the more so, of course, in one's own home. Jane's interest, insights, support, and encouragement created the space every writer needs to engage with their work. I am grateful.

Authors typically add to their acknowledgements the qualifier that the shortcomings of their work are exclusively their own. They are. Despite their best efforts, my advisors have not always convinced me to adopt the many fine additions and modifications they have suggested.

About the Author

A Mind of Its Own?
Chat GPT and the Surprising World of the New AI

Terry M. Savage, Ph.D.

For thirty-five years, Terry M. Savage taught philosophy and interdisciplinary humanities courses at the University of New Hampshire. He did graduate studies with the Committee on Social Thought at the University of Chicago and earned his Ph.D. (Philosophy) at Boston University.

A winner of his university's highest teaching award, Dr. Savage has a long-standing interest in the educational uses of computers. He was an early adopter of multimedia technology and in 2009 co-authored, with Karla E. Vogel, *An Introduction to Digital Multimedia*.

His interest in artificial intelligence spans many years of teaching related to the history, technical development, and social impact of AI technology. *A Mind of Its Own?* is his exploration of the nature and potential of the surprising powers of ChatGPT and other large language models.

It is offered in the hope that it will contribute to a better understanding of a new technology that appears destined to change all our lives in many ways.

Made in United States
North Haven, CT
27 January 2025

64946585R00129